AF624295

INDUSTRIAL FAMILIES IN INDIA

INDUSTRIAL FAMILIES IN INDIA

An Enquiry into the Nature of Their Entrepreneurship

H. S. VERMA

CONCEPT PUBLISHING COMPANY, NEW DELHI

First Published 1987

© H. S. Verma 1985
Harnam Singh Verma (*b*. 1942-

Published by

Naurang Rai
Concept Publishing Company
H-13, Bali Nagar
NEW DELHI-110015 (India)

Printed in India by

Printway India
at S. Narayan and Sons
7117/18, Pahari Dhiraj,
DELHI-110006

To
Neelima and our two daughters,
Neera and Neeta

PREFACE

Most analyses of industrial entrepreneurship in India are concerned mainly with the strategic response of the entrepreneurs in their economic units and occasionally with the negotiation of the external business environment. One major cluster of variables which also plays a significant role in shaping the performance of entrepreneurship, especially in the large-scale manufacturing sector, is the structure and functioning of the domestic institution of 'family'. This has not received the adequate attention of the research fraternity. My objective in writing this book was specifically to examine the impact of structure and functioning of industrial families on the management of economic enterprises. I would have been methodologically correct if I had confined myself only to the analysis of the impact of the structure and functioning of industrial families on such enterprises. However, I have carried it further to include the control of internal environment of the economic units and negotiation of the external business environment, in a more broad-based effort also to gauge the relative capacity of the three clusters of variables in explaining entrepreneurial performance. I also felt that measurement of entrepreneurial performance only on the basis of results obtained by the economic units did not amount to a correct delineation of entrepreneurship. Consequently, I decided to use own wealth, managed wealth and total wealth of my sample families as three separate determinants of entrepreneurial performance. Since this study was historical in its approach and sensitive in terms of data requirements, the task which I set before me was quite formidable.

In completing this study first as a Ph.D. dissertation in Sociology submitted to the Gujarat University, Ahmedabad, and subsequently as a book in the present form to a wider readership,

I have received encouragement, co-operation and help from a large variety of individuals. Surely it is not an easy task to repay this debt. To the extent it is possible I am, therefore, recording its nature and intensity.

My debt to my supervisor, Professor V.R. Gaikwad, Indian Institute of Management, Ahmedabad (IIMA), is indeed very considerable. It was he who suggested the theme of the enquiry, gave me freedom and encouragement to pursue a promising line of investigation, guided me in the right direction in my moments of doubt, and offered me a wide variety of resources to complete the study. But for his support and guidance it is inconceivable that the study could have been completed at all. In clarifying many of my doubts I have also benefited a great deal from several rounds of discussions with Professors D. Tripathi and P.N. Misra (IIMA) ; Dr. Radhika Ramasubban (Bombay University) ; Ranjit Roy (Resident Editor, *The Economic Times*, New Delhi) ; Krishna Raj (Editor, *Economic and Political Weekly*, Bombay) ; Bharat Jhunjhunwala (formerly at IIM, Bangalore) ; and Ashim Chowdhury (North Bengal University, Siliguri).

For an intensely involved researcher, recasting of the dissertation to make a presentable book posed a major hurdle. Fortunately for me, the detailed comments received from Professors A.R. Desai (Bombay) and R.K. Mukherjee (Calcutta) as examiners of my doctoral dissertation dealt almost exclusively with this aspect. The task of reorganizing the material consequently became much easier and more purposeful.

Over 350 city-based local resource persons helped me in data collection on sensitive aspects of the study. In their interest, unfortunately, I am duty-bound to withhold their names. Once the initial bridge had been built with the help of a kingpin, most of my respondents gave me their time—reluctantly first but ungrudgingly later—and extended all help and co-operation. In a few cases they went out of their way to help me out. But for these courtesies the study would have certainly taken much longer to complete than it eventually did. I received excellent computational help mainly from J.P. Misra and Sapan K. Ghosh and, to a lesser extent, A. Joshi, S.D. Rai and G.D. Patel assisted with the layout of the illustrations.

The Vikram Sarabhai Library of the IIMA and its ever helpful duo, Pandya and Trivedi, provided excellent library support.

This was occasionally supplemented with help from the libraries of the Administrative Staff College of India, Hyderabad; Universities of Bombay and Lucknow; Sardar Patel Institute of Economic and Social Research, Ahmedabad; the Tata Institute of Social Sciences, Bombay; the National Library, Calcutta; and Moti Mahal, Lucknow. My respondents themselves of course provided me the single largest source of referencing material including unpublished and private documents.

The study commenced in July 1976 at Ahmedabad when I was working at the IIMA. I shifted to the Giri Institute of Development Studies, Lucknow in September 1977. This necessitated constant co-ordination between the two types of institutions, my supervisor Professor Gaikwad and I functioned. K.M. Varunny, as Professor Gaikwad's secretary, provided the link and helped to have my problems solved. M.S.K. Rao handled the typing of the manuscript.

Prosecution of a study of such wide coverage required considerable financial commitment. Most of the expenses had to be met from not too bountiful resources of my own. In sharing them, my wife Neelima and our two daughters Neera and Neeta, had to forego what were their legitimate and rightful individual shares. My financial liabilities were, however, somewhat reduced owing to the grants sanctioned by the Indian Council of Social Science Research, the Sardar Patel Institute of Economic and Social Research, Ahmedabad, and the institutional support extended by Dr. S. Paul, the then Director, IIMA, and Dr. T.S. Papola, Director, Giri Institute of Development Studies, Lucknow.

While this study was under way, I had to go through tremendous emotional turmoil because of tragic events in my family. In facing the trauma and simultaneously completing this study, I received love, compassion and understanding from my family members demonstrating in the process the jointness of this enterprise. As I argue in the text of this study, its completion was as much shaped by events in the internal structure of my family as developments outside its domain. From my close peer group I received constant solace and help—from Ilyas, Kapil, Ashvin, Virendra, late Waheed, Dhirandra Narain, Ram Dhani and Sarvajit—when it was needed most.

It is my firm conviction that the book has gained much from each of the sources mentioned. However, for its inadequacies I am alone responsible.

50, Gulistan Colony, Varanasi Road,
Lucknow-226001

H.S. VERMA

CONTENTS

LIST OF FIGURES

LIST OF TABLES

1

INTRODUCTION

ENTREPRENEURSHIP as an abstract process is directly related to the larger processes of development and underdevelopment in any society. How a country has interacted with the global forces and shaped its development goals and programmes could be easily gauged from the nature of entrepreneurship seen in its society. Though India is fairly well endowed with natural resources, entrepreneurship in the industrial field has displayed a very narrow, regional and communal base (Bagchi, 1972, p. 157; Buchanan, 1934, p. 145; Chowdhuri, 1975, pp. 108-144; Jha; 1963, pp. 102-168; Pavlov, 1964; Timberg, 1978, p. 18). As a result, industrial development through private efforts has been slow and uneven.

As an area of research entrepreneurship has attracted the attention of a large number of scholars who have analysed its various aspects (Timberg, 1978, p. 18). Thematically these studies can broadly be grouped into six categories :

1. *Theoretical and comparative* : Examining the concept of entrepreneurship; mono-factor explanations of its causation or inhibition; successful or unsuccessful operation at the individual, community and societal levels; comparison with the entrepreneurial profiles in other countries.

2. *General studies on business communities* : Analysing the general character of business communities from different regions.

3. *Specific community studies* : Depicting community characteristics and explaining their success or failure.

4. *Localized studies* : Describing the town-or city-specific ethos in entrepreneurial styles.

5. *Biographies and histories of individual firms* : Listing the sequential development of business houses or firms.

6. *Economic history* : Chronicling the historical evolution of industrial and commercial institutions, sometimes linking it up with the processes of development and underdevelopment; concentration of private economic power and its juxtaposition with policies pursued by the Indian government, and the role of multinationals.

Taken together these studies do provide insights into different aspects of Indian entrepreneurship. However, they suffer from the several inadequacies.

Taken singly none of them fully and satisfactorily explains the phenomenon of entrepreneurship (Tripathi, 1976, p. 2). The monofactor approach tends to concern itself with either the socio-economic, cultural and political environment in which the entrepreneur operated, or personal, community and societal characteristics that explain his success or failure (Alford, 1977, p. 116).

Quite often these studies fail to make a distinction between different dimensions of the phenomenon :

(a) the individual entrepreneur, his background, style and effectiveness ;
(b) the nature of his enterprise ; and
(c) his response vis-a-vis operational environments (Alford, 1977, p. 116).

Each of these dimensions is an independent issue and has its parallel significance, and yet most of the studies tend to confuse them (Timberg, 1978, p. 18).

Since most of the studies follow the Weberian or neo-Weberian framework originating in the west, their operational methodology use the same values and theoretical bias, and consequently examine entrepreneurship in a linear fashion (Epstein, 1964 ; Javillonar and Peters, 1973, pp. 314-18). This tainted methodology has resulted in inappropriate measurement of empirical facts and in reaching biased conclusions.

The dominant tradition in the studies has been to treat entrepreneurship as a process of challenge and response. In this process the entrepreneur acts as a central dynamic agent seeking to achieve some combination of profit maximization and growth of his enterprise. The precise degree of entrepreneurship depends upon his responsiveness to wider social and cultural conditions. In this approach factors such as markets and technology are seen as being given. In evaluating the choices open to the entrepreneur, the relative contribution of entrepreneurship as against other factors is seldom included in an enterprise's performance. Thus a great many studies on entrepreneurship have been concerned with the entrepreneur *per se* and not with his business performance (Alford, 1977, p. 116).

The studies have also contributed, directly and indirectly, to the emergence of the following myths :

1. That certain Indian communities have played a leading part in India's industrialization because they are disadvantaged minority groups (Lewis, 1960). That they have values that differ from the dominant Hindu tradition (Kennedy, 1962 ; Hoselitz, 1960 ; McClleland, 1961). That there are few intra-community differences in characteristics and performance in business and industry. The facts are otherwise. Excepting the Parsees, there are important differences in most of these business communities. Ancestors of all Marwaris, for example, belonged to Marwar ; beyond that there are religious and caste differences among them. The same is true in the case of Gujaratis.

2. That the suppliers of capital were prime movers of the process of economic growth. That these were businessmen and that the feudal and pre-capitalist remnants of upper income groups, which were preserved by the imperial power, usually did not play a constructive part in the production process. Bagchi (1972) has not only identified these misconceptions but also provided convincing evidence to rebut them for the period 1900-39.

3. That foreign enterprises provided the only dynamic element in the Indian economy through their ability to take risk and innovate during British rule. That the multinationals have continued the same role after the independence and Indian business has seldom operated on its own strength (Negandhi and Prasad, 1971 ; England, et. al., 1974).

4. That Indian business could be stratified on size alone. That concentration of economic power in the private hands could be curbed by putting controls over their further growth

(Monopolies Enquiry Commission, 1963; Industrial Licensing Policy Enquiry Committee, 1969). Chowdhuri (1975) and Tyabji (1977) indicate that this could also be done on the basis of geographical coverage and legal forms.

Compared to the earlier studies, the conceptual distinctiveness of the present study lies in its treatment of entrepreneurship as a historical, non-linear process and a family rather than an individual phenomenon. It is the contention of the study that entrepreneurship in the Indian large-scale manufacturing sector is quite a diffused phenomenon, one vested in the collective authority structure of the joint family headed by a *karta.* While it is true that the *karta* does indeed play a very leading part in strategic decision making, it is also a fact that he is not the only one involved in decision making. In point of fact he is only a senior leader of a team often consisting of equally senior members. The focus in the present study, therefore, is on the authority structure in a joint family rather than on discovering charismatic qualities in one individual. The unit of analysis is the industrial family as against an individual.

In evaluating the business performance of the families the study simultaneously considers the three sub-systems of the business environment: the structure of family, management structure of the industrial units and external business environment. As a result the analysis covers the structure and content of family entrepreneurship rather than the individual entrepreneurial qualities *per se.*

Following this approach, Chapter 2 first reviews mono-factor attempts—cultural, psychological, sociological and environmental—weighing their explanatory adequacies. It then examines how entrepreneurship is a historical, non-linear process and why it should be judged on the basis of the performance of entrepreneurial tasks. Next, the rationale behind analysing entrepreneurship as a family phenomenon in India is outlined. It is argued that (i) strategic decision making in the industrial families is of a diffused nature; (ii) the basic unit for individual identification and orientation in India is the family; (iii) the socio-cultural practices and personal laws overlap the laws dealing with the regulation of industrial enterprises and encourage management of industrial enterprises within the legally flexible ambit of the family; (iv) structurally, the Indian family has encompassed a variety of activities and organizations controlled by its members and (v) the family represents a unit of class theory.

The Indian joint family, as a field of research interest, has received the attention of many researchers Indian and foreign. Yet, a great deal of confusion exists about the concept of family, its typologies, and its relationship with the larger societal processes such as industrialization, urbanization and development (Dube, 1973, pp. 233-366 ; Madan, 1976, pp. 212-13 ; Shah, 1964, pp. 1-36).

Chapter 3 first reviews the three studies that deal with the interacting relationship between family and management of enterprises. A conceptual framework of the study is outlined next by specifying functioning of the family as a corporate entity allocating or reallocating resources and the manner in which the demographic changes in the family and critical events affect the authority structure and consequent operationalization of the business environment. Based on this discussion the specific objectives of the study are evolved. These are :

(a) to study the demographic and authority structure of industrial families on a historical continuum ;
(b) to analyse the impact of changes in the internal structure of the families on the management of enterprises ; and
(c) to identify, on the basis of the findings, areas of excellence in the management of the internal structure of the industrial family and their economic units which could be recommended for adoption by any professionally managed enterprise.

The specific hypotheses generated for empirical testing are :

1. Changes in the demographic structure of the family are likely to change the authority structure of the family.

2. Changes in the demographic structure are likely to change the deployment of family members in the economic enterprises.

3. The nature of the authority structure is likely to determine the presence or absence of turbulence in the family.

4. Business performance is likely to be dependent on the effectiveness of strategies employed to combat familial turbulence, and to control internal and external business environments.

Chapter 4 outlines the thematic and historical scope and geographical coverage of the study. Research design, sample, tools and methods of data collection, collation, restructuring and recasting, including the problems encountered and the strategy employed to solve them are described. Operational definitions of the concepts used are indicated. Limitations of the study are also listed. A pilot study was conducted to evolve the universe of 96 families. From the universe a sample of 33 industrial families was selected on a random basis.

Analysis of the data is done in two stages. In the first stage both the clusters of variables, independent and dependent, are independently presented. The relationship between the two sets is shown in the second stage. The first stage consists of Chapter 5-9 which depict the position of different variables at different points of time. Chapter 10 comprises the second stage where the relational matrix of the dependent and independent variables are shown in an effort to test the generated hypotheses. The demographic structure of the families is examined in Chapter 5 in terms of characteristics indicating categories of jointness, size and composition. Changes in it are presented in terms of births, deaths, marriages and adulthoods reached. Management structure of various types and changes in them are discussed in terms of size and composition in the family and enterprise unit levels. Authority and power structure of the industrial families decides the nature of entrepreneurship. Functioning of the authority structure in the domestic sphere is examined in Chapter 6 by considering the educational attainments of different members, professional training or education, familial background of the grooms for daughters and brides for sons of industrial families, occupational profile in the pre- and post-industrial period, deployment of familial economic resources, presence of turbulence and nature of strategies used to combat turbulence.

The operation of the authority structure in the internal business environment is viewed in Chapter 8 covering areas of industrial units (total and new units, expansions, diversifications, products covered), the workers, shareholders and the support systems. The external business environment is analysed in Chapter 9 vis-a-vis the state, the suppliers and dealers, consumers, and multinationals and other competitors. These three sub-systems provide the content of the entrepreneurship displayed by different families at different points of time. Four main and three related hypotheses dealing with the relationships of (i) demographic structure and changes in it with the management structure and changes in it, (ii) authority structure with the presence of turbulence in the family and effectiveness of strategies employed to combat them and (iii) the relationship of effectiveness of strategies employed (in combating familial turbulence and in controlling the business environment) with business performance are tested empirically in Chapter 10. Extending the analysis still further, it also indicates the relative capacities of these variables in explaining business performance.

Chapter 11 identifies positive and negative aspects of management of the industrial families and their economic units, to indicate areas for emulation and correction, respectively. The role played by family in the formation of capitalist class, its composition, inter- and intra-class relations is examined in the background of data generated by the study. Validity of myths in vogue about the role of traditional organizations such as family vis-a-vis complex organizations are juxtaposed with the empirical facts reported in the study to weigh their validity. Finally, priority areas for future research in organizational sociology are identified.

2

APPROACHES TO THE STUDY OF ENTREPRENEURSHIP

THE scholarly tradition of studying entrepreneurship is nearly half a century old. Yet, significant differences in opinions continue to persist about its meaning, causation and operation on the one hand and the approaches and methodologies used to study it on the other. The understanding of the concept itself is a result of several different stages of definition, quite often not properly differentiated from one another (Timberg, 1978, p. 18). Very few researchers appear to have recognized the fact that an entrepreneur performs a large number of roles sequentially in relation to his enterprises; specific institutions of the society forming part of his socio-eco-political environment, and the global forces affecting the process of social transformation.

In this chapter, mono-factor attempts in describing the phenomenon of entrepreneurship is first reviewed.[1] Their explanatory inadequacy is then pointed out. How entrepreneurship is considered to be a process and should be judged on the basis of performance of entrepreneurial tasks is explained thereafter. Finally, the approach to the study of entrepreneurship in India as used in the present study is outlined.

A concurrent feature of most research studies on entrepreneurship has been to attempt mono-factor explanations of various dimensions of entrepreneurship. Factors identified by such studies have been considered under a generally agreeable grouping to review them in the analysis that follows :

CULTURAL FACTORS

Two sets of scholars trace bases of entrepreneurship in ethical values. The first of these is best typified by Max Weber (1930), who found a positive relationship between the Protestant ethic and industrial capitalism.[2] The second set is not much concerned about why industrial capitalism did not develop in certain societies. These scholars concentrate their efforts in identifying certain attributes which, according to them, contribute to the growth of entrepreneurship (Hoselitz, 1960 ; Kennedy, 1965 ; McClleland, 1961). Weber (1930) postulated that entrepreneurial growth was dependent upon ethical value system of the society concerned, that rapid industrial growth depended upon a rationalised technology, acquisition of money and its rational use of productivity, and multiplication of money and that entrepreneurship depended upon rational attitude towards action.

Weber then proceeds to provide positive and negative illustrations of his theoretical postulates. Thus, he discovers a correlation between the Protestant ethic and the spirit of capitalism. Hindu, Jain and Jewish values, he finds, obstruct development of industrial capitalism (Weber, 1960, pp. 193-204). Weberian thesis on Hindu values is also partially supported by Sarkar (1937, pp. 6, 634-36) and Goheen, et. al. (1958, pp. 1-12). Sarkar postulates a substantive identity between inherent 'rationalism' of Christianity and the unbroken 'materialistic' tradition of Hinduism. Goheen, et. al. allege that the 'contemplative' behaviour patterns of Indians obstruct the country's economic development.

There is a neo-Weberian group of scholars which restates the Weberian thesis without its historical trappings. This group comprises Eisenstadt (1968), Kennedy (1965), Peacock (1969) and Yale (1954, pp. 339-64). It concludes that entrepreneurship develops rapidly in those societies where ethical values provide independent capacities of decision making. McClleland (1961, p. 178) explains differential development of societies in terms of their willingness to break with the institutional survivals of an earlier period and acceptance of social and technical innovations. He emphasizes the role of religious doctrines which stress individual responsibility, and thus help in fostering the self-reliance necessary for entrepreneurship. Hoselitz (1960, pp. 151-55) similarly concludes that entrepreneurship could develop only in a society in which cultural norms permit variability in the choice of paths of life, and in which the relevant processes of socialization are not completely standardized.

A critical look at the Weberian thesis must separate its application to western societies and India. In so far as its suitability to western societies is concerned, it has been questioned on several grounds. Empirical evidence is against his observations on industrialization owing to Protestant ethic. Samuelson (1961), for example, states that capitalism also developed in those societies where Protestant ethic was not present. Some of the ethical values identified by Weber and his followers might have different implications for different societies. Nakane (1970) for instance notes that collectivism (as against individualism), much greater than the one seen in Hindu society, has not hampered industrial development in Japan. Smith (1965) and Gusfield (1967) point out that traditional societies change in their own ways, not necessarily following the western model, ethics and values.

Weberian theory in so far as it applies to Hinduism and Indian society has been criticized by western and Indian scholars on the grounds that Weber's understanding of Hindu value system is faulty, and that empirical evidence is against his observations on industrialization in India.

Critical response of western and Indian scholars to Weberian postulations is in terms of assessing the rationality content of Hindu tradition. As Gupta (1974, pp. 13-43) notes, this forces them to believe that Hinduism can be generalized and defined in terms of Christianity's 'irrational' attributes, that evolutionary modernity in India can occur under the benevolent impact of the west since this involves only imitation and adaptation ; and that India's future modernization can be assessed on the criteria laid down by an already modern western society. As a consequence, they attempt a typical Hindu world view which can visualize changes only terms of adjustment, imitation and alliance. Thus, even though on surface they appear to be quite opposed to Weber, in reality they become instrumental in promoting an inappropriate anology for Hinduism's encounter with industrialism and development.

PSYCHOLOGICAL FACTORS

Among the psychological factors considered to be affecting entrepreneurship, industrialization and economic development are (i) status marginality and minority group morale, (ii) need for achievement motive, (iii) empathy and cosmopoliteness and (iv) spirit of modernity.

STATUS MARGINALITY AND MINORITY GROUP MORALE

Those advancing such view fervently believe that entrepreneurial growth has something to do with the minority group morale. Because the minority group is otherwise handicapped, nurses a feeling of powerlessness, impotency and lack of an identity, and feels that it is not given its due place in the mainstream of society, it tries to compensate by doing well in industry.

Before Hagen elaborated his theory of 'withdrawal of status-respect', Park (1928) and Stonequest (1937) had attributed causation of technological innovations to status marginality. Hagen (1962, p. 113) later developed the theory that the disadvantaged minority group is mostly the source of entrepreneurship. Taking Japanese history as his starting point he traced Japan's quick development (in relation to other non-western societies excepting the USSR) to two historical differences, (i) the Japanese were free from 'colonial disruption', and (ii) they experienced long, continued withdrawal of expected status from important groups (i.e., Samurai) in their society. This drove them to retreatism, alienated from traditional values, and increased productivity. Much later also, discussing the case of Pakistani Memons, Hagen (1962 a, p. 61) explains the innovative behaviour of groups in traditional societies in only terms of his old theory.

Hagen (1971, pp. 123-38) regards three requirements essential for the transition to economic growth, (i) fairly widespread creativity—problem solving ability and a tendency to use it, (ii) attitude towards manual-technical labour and the physical world and (iii) channelization of creative energies into innovation in the technology of production rather than in the technology of art, war, philosophy, politics or other fields.

Frayer (cited in Pavlov, 1964, p. 89) and Lewis (1960) emphasize that the Parsees could give a lead in Indian entrepreneurship because traditionally they enjoyed prestige which was denied to them later on. Their desire to secure back their lost prestige found expression in their successful entrepreneurship. Even McClleland (1961, pp. 7 and 15) tacitly supports this viewpoint. For example, he attributes the greater commercialization of the Jains in India to their awareness of being a minority and a superior group. This, according to him, generates a high degree of need achievement in them. Arguing in the same vein, Hazelhurst (1968, p. 176) feels that truly innovating commercial and investment ideology occurs only within the traditional merchant population whose normal identity has been disrupted by forced migration following partition. The importance of the attitude toward occupation, the role expectations held by sanctioning groups, and the operational requirements in determining the success of businessmen has also been stressed by Cochran (1971, pp. 95-108).

It is true that some—and this might well be an insignificant section—among the minority and disadvantaged groups have indeed acted on the lines indicated by Hagen and McClleland. Historical evidence, however, makes it difficult to sustain the generalization for all minorities and all sections among them. Minority and disadvantaged groups in the U.K. (Catholics) and the United States (Negroes) have, for example, not supplied many entrepreneurs. The same is true of the Muslims, Christians, Harijans and Scheduled Tribes in India.[3] Even the feeling of superiority, which McClleland links with a high degree of need achievement motive, does not appear to be working for the dominant Brahmins and Rajputs in India. Perhaps, there is more substance in the argument advanced by Epstein (1964, pp. 162-65) that entrepreneurship among the disadvantaged group is dependent upon community support. Wherever this support is lacking, the group has to be propped up by the state apparatus or else its members do not do well at all.

NEED FOR ACHIEVEMENT MOTIVE

McClleland (1961) is the chief exponent of the view that tries to explain entrepreneurship in terms of role theory. Following him, entrepreneurial growth is explained by the need for achievement motivation (N.Ach.). N.Ach. is a major determinant of industrial and economic growth.

McClleland measured N.Ach. by analysing the entrepreneurial growth of different communities and societies on the basis of his interpretation of their ideologies reflected in literature, art, history and religion. According to him, the Jains and Parsees progressed economically owing to a high degree of their N.Ach. as a result of their child-rearing practices (McClleland, 1961, pp. 7-15 ; McClleland and Winter, 1969). Later on he expanded his set of motives to include N. Power, N. Affiliation to his original N.Ach. (McClleland 1971, 1972, 1975).

There is no substantial difference in the views of McClleland, Weber and Hagen. In ideological terms, what is N.Ach. to McClleland is 'rationality' or 'economic-rationality' to Weber, and 'increased creativity' to Hagen. All of them relate legends, child-rearing practices and ideologies with entrepreneurship, but in different ways. Whereas McClleland directly relates N. Ach. with entrepreneurship, Weber and Hagen relate only those social factors with entrepreneurship which generate or are the sources of N. Ach.

Despite its appeal, McClleland's viewpoint is far from perfect. Redlich (1963, p. 19), for example, has called McClleland's interpretation of causation of entrepreneurship as 'incomplete' since it examines only the interaction between the entrepreneurs and the community and/or society. The single most criticized aspect of McClleland's theory is its implied assumption of social homogeneity in terms of entrepreneurship.[4] Entrepreneurship demands differential personal capacities under different economic and cultural conditions (Nandy, 1973, pp. M-98-M-106 and Owens and Nandy, 1978).

EMPATHY AND COSMOPOLITENESS

Empathy, as defined by Lerner (1964), is the ability to relate oneself to a changing environment, or the ability to introject other roles and project oneself into different roles. He considers empathy a very crucial variable in the modernization process. According to him, it helps the individual to adjust and stabilize himself in a changing environment, visualize various possibilities, and plan to act accordingly. In Lerner's scheme, empathy itself is sharpened with increased mass-media exposure, higher literacy, political participation and urbanization.

Cosmopoliteness is considered to be natural extension of empathic abilities and refers to one's orientation to the larger society that lies beyond one's immediate surroundings. Cosmopoliteness, following Merton (1961) and Loomis (1962), indicates awareness of alternative norms of behaviour and the general trend of national events which also influences the local community. It is supposed to induce acceptance of new ideas and practices.

Lerner (1964) and Rogers (1964) report that empathy and cosmopoliteness are positively correlated with innovativeness. However, the meaning assigned by them to innovativeness is not exactly what Schumpeter assigns to entrepreneurship. Nevertheless, it is generally agreed that innovativeness and entrepreneurship are positively correlated.

SPIRIT OF MODERNITY

Arora (1968, pp. 85-120) and Dutta (1971), in two separate critical evaluations of theories on modernization, note a conscious effort to emphasize certain western values—social, economic, political and even religious—to be the spirit of modernity. At the root of this theorization lies what has come to be identified as the 'value-convergence' hypothesis. The proponents of the value convergence hypothesis not only treat the western values to be the best, they also consider the phenomenon of increasing global consequence. There is an explicit assertion in it that it involes similar structural changes wherever it occurs (Parsons, 1964 ; Feldman and Moore, 1965 ; Meyer, et al. 1975). It is assumed that, from the spread of industrialism, world's societies are now becoming more alike in some respects than they were at some earlier point of time (Schnore, 1961). Lastly and more importantly, there is some convergence of values across the cultures as a result of this development (Weinberg, 1969 ; Meyer, et al. 1975 ; Inkeles and Smith, 1974).

Most of the 'value-convergence' attempts have centred on constructing indices of the syndrome itself and the question of what it means remains open (Fliegel, 1976, pp. 431-51). Inkeles and Smith (1974), for example, have constructed an 'overall modernity' (OM) scale consisting of 24 different themes that McClleland (1976, pp. 161-65) finds bristling with conceptual confusion and methodological inadequacies. Its composition differs in six countries where it was empirically validated. Assuming that such indices are clear and methodologically sound, there is conflicting empirical evidence to prove non-convergence of values, beliefs and attitudes (Portes, 1973 ; Aronoff and Crano, 1975 ; Fligel, 1976).

SOCIOLOGICAL FACTORS

A large number of social scientists have identified social institutions that are either considered helpful in fostering entrepreneurship or recognized as hindrances in its development. Institutions such as caste, religion and joint family are alleged to be obstacles to entrepreneurial growth. Maine (1871), for example, considers caste and rigid contractual basis of social life, the joint family, the related institutions of customary law, and the crippling restrictions under which women have traditionally been forced to live in Hindu society to be manifestations of backwardness. Davis (cited in Goode, 1963, p. 205) holds the caste system responsible for limiting competition, and of hampering the development of large-scale co-operative enterprises. Singer (1972, p. 276) notes that the most frequently alleged obstacles to modernization in India are the ritualism of the caste system and the doctrine of fate (*karma*), rebirth (*punarjanma*), duty (*dharma*) and salvation (*moksha*) which are supposed to constitute a theodicy for the social system's moral and metaphysical justification. Two well-known exponents of this view are Max Weber (1960) and Kapp (1963). The Hindu joint family has similarly been recognized as an obstacle to and incompatible with economic development by an impressive number of scholars (Wirth, 1938 ; Weber, 1960 ; Kapadia, 1959 ; Wolf, 1955 ; Parsons, 1943, 1949 ; Parsons and Bales, 1963 ; Nimkoff, 1960 ; Nimkoff and Middleton, 1960 ; Ogburn and Nimkoff, 1950 ; Levy, 1966 and Goode, 1963, 1965).

It has been alleged that industrial pooling of resources and control by the *karta* in the Hindu joint family kills initiative. The authority structure discouraged collective decision making on rational grounds. Innovativeness does not operate. Emphasis on hereditary occupation discourages the mobility necessary for increased competition, the acquisition of new skills, and entry into new occupations. Physical and mental development of children is neglected, and wasteful expediture on rituals limits savings (Anstey, 1952 ; Davis, 1948 ; Misra, 1962 ; Kapp, 1963 ; Gore, 1963, 1968 ; Davis in Goode, 1963, p. 205). Benfield (1958, p. 7) after analysing an Italian village similarly concludes that 'amoral familism' acts as a limiting factor in the way of economic development in most of the world. At the other extreme, rationalism, Protestant ethic (Weber, 1960) and nuclear families (Goode, 1965) are cited as bastions of economic development.

That Hindu religion has not hampered industrial entrepreneurship has theoretically been controverted by a large number of writers earlier cited. Cohen (1973), Timberg (1969, 1978) and Singer (1972, 1973) provide some interesting empirical evidence. The 'functional-fit' discovered by Goode (1965) in the conjugal family and the modern industrial system has been countered on the following grounds :

1. Conjugal families have been present in the pre-industrial west (Laslett, 1977 ; Laslett and Wall, 1972).

2. Great similarities exist between pre-industrial and post-industrial families in the west (Furstenberg, 1966 ; Lanz, et al. 1968, 1973, 1975, 1977).

3. Certain strains, which are often attributed to the advent of industrialization, were evident in pre-industrial families (Furstenberg, 1966 ; Lanz, et al. 1968, 1973, 1975, 1977).

4. The Hindu joint family has in fact promoted industrial development (Gaikwad and Tripathy, 1970 ; Srinivas, et al. 1966 ; Fox, 1973 ; Singer, 1972 ; Ramu, 1973).

5. The joint family as such has been helpful in entrepreneurial activities in other non-western societies as well (Litwak, 1963, 1975 ; Khalif and Shawayri, 1966 ; Benedict 1968 ; Nafzigar, 1969 ; Sussman, 1970 ; Harris, 1971 ; Marris, 1975).

There is conflicting empirical evidence about the role played by another traditional collectivity, the community, in entrepreneurial growth. One group of researchers holds that community characteristics help in the better performance of modern economic activities ; another group feels just the reverse. From the first group, Srinivas, et al's (1966) Ramgarhias at Okhla, K.N. Sharma's (1967) small-scale entrepreneurs in and around Kanpur, Owen and Nandy's (1978) Mahisas in Calcutta, Glass-burner's (1962) entrepreneurs in Indonesia and Papanek's (1973, pp. 61-106) Memons in Pakistan draw immense help from their communities. Among linguistic and regional but multicaste communities such as the Marwaris and Gujaratis, Brimmer (1955, pp. 551-76) ; Lamb (1955, 1956) ; Gadgil (1951, 1959) ; Pavlov

(1964), Hazelhurst (1968) and Timberg (1969, 1971, 1973, 1978) discover attributes that have helped their performance in the industrial field. On the other hand, Pathak (1970, 1973), Singh (cited in Nafzigar, 1973, pp. 287-317) and Mines (1972, 1973) from the second group find community attributes inhibiting entrepreneurial performance.

ENVIRONMENT AND SUPPORT SYSTEMS

Environment and support systems include the following factors :

(a) New Ideas and Opportunities ;
(b) Socio-political System ;
(c) Regional Industrial Climate and Ethos ; and
(d) Geographical Factors.

New ideas and opportunities

Tripathi (1971, pp. M-59-65) notes that between the pioneering Parsee and Hindu entrepreneurs the common factor has been exposure to new ideas and values. K.N. Sharma (1969, pp. 474-93) on the other hand, attributes the differential response of social groups to the opportunities provided by the commitment of the political and cultural systems to industrialization. Sharma argues that the Parsee value system did not preclude contacts with the British, whereas the Hindus were less close to the British because of their religious values. The Muslims were hostile. Tripathi and Sharma differ on what causes exposure to new groups, ideas and opportunities. Whereas Tripathi emphasizes the role of education in doing so, Sharma stresses the contribution made by the traditional collectivities based on religion, region and caste.

Socio-political system

The spirit of entrepreneurship exists in every society. However, the encouragement given to its flowering by the support systems, among other things, also determines its success. How have the socio-political systems in different societies provided the support ? On the basis of a historical study Building (cited in Yale, 1954, pp. 339-64) finds political structure in addition to societal hurdles such as the concentration of power and property in the aristocracy, a decisive factor in the entrepreneurial growth of France and the Soviet Union. Hoselitz (1955, p. 291) suggests that the social structure of France did not provide sufficient incentive and security to the entrepreneurs. Entrepreneurship flourished in Japan because both the social structure and the government provided it adequate support and incentives (Hierschmer, 1964 ; Holt and Turner, 1966 ; Vepa, 1967 ; Ward and Rustow, 1967).

In India, the effectiveness of the support systems since the British period has been critically reviewed by many. An overwhelming majority of historians blames the British for agonisingly slow and unbalanced industrial growth, although there are others who suggest that the British rule did, on balance, positively contribute to India's industrial growth.[5]

In the post-independence period Medhora (1965, pp. 558-86) attributes late entrepreneurial growth not so much to lack of entrepreneurial motivation but to the non-commitment of the political structure. In the more recent studies the impact of public policies on entrepreneurial growth in the large-scale and small-scale industrial sectors has been viewed differently by various researchers. On the one hand there is a group that recognizes their positive contributions (Sharma, 1965, pp. 355-68 ; Mishra, 1966, pp. 133-40). In the small-scale sector this is particularly evident in Maharashtra and Gujarat where the support systems are well organized and function fairly well. There are others who are critical of the policies pursued by the government (Chowdhury, 1975 ; Mehta, 1952, 1961 ; Mahnot, 1962 ; Hazari, 1966 ; Kothari, 1967 and Rungta, 1970).

Regional industrial climate and ethos

Pandit (1957, pp. 282-86, 315-17) advocates the role played by regional industrial climate in entrepreneurial growth. She challenges the viewpoints that emphasize the contribution of religious values and group community characteristics in industrial entrepreneurship. In an interesting study of Bombay and Ahmedabad regions, she explains the dominance of Gujaratis and the Marwaris in terms of regional industrial climate in these two regions. Because of variation in the climate, the Gujaratis responded differently in these two places: the Bombay entrepreneurs were pioneers ; those of Ahmedabad, in the words of Spodek (1969, pp. M-23-31), turned out to be 'pioneer imitators'. The entrepreneurs in Ahmedabad did not diversify. Pavlov (1964) and Timberg (1969, 1971, 1973, 1978) assign better suitability to regional climates as the main reason for the successful exploitation by a few communities.

At the other end of the continuum are the primitive communities with their distinctive cultural and socio-economic characteristics. They too have entrepreneurs who experiment, innovate and take risk. However, as Moulik (1975, pp. 54-61) and Bogaert (1975, pp. 62-75) point out, they are different in their response, operational style and system orientation. They face an altogether different set of handicaps.

Geographical factors

Topography, physical resources and climate, among others, are considered key geographical factors that have played important roles in industrialization and economic development. Kamarck (1976) considers climate to be the most crucial among all of these because it influences agriculture, mineral discovery and man's health (through diseases). Climate, however, cannot be always blamed for poor societal record on the economic front since it is known that with the technological advances made so far, it can be modified, adapted and even controlled for habitation and production purposes. Nor can the poor physical resource endowment be held responsible for the underdevelopment of nations. Human history provides enough examples of nations with poor resource endowment taking rapid strides in the industrial field.

EXPLANATORY ADEQUACY OF IDENTIFIED FACTORS

Having discussed the factors that have been said to foster entrepreneurship at individual, societal or global levels or alternately, to hinder its development, their critical assessment may now be considered. It is more than evident that taken singly, none of them explains the phenomenon of entrepreneurship fully and satisfactorily (Tripathi, 1976). This single factor approach tends to concern itself with either the socio-economic-cultural-political environment in which the entrepreneur enters and operates or personal, community and societal characteristics that explain his success (Alford, 1977, p. 116).

Quite often these studies fail to make a distinction between different dimensions of the phenomenon : the individual entrepreneur (i.e., his background, style, effectiveness) ; the nature of his enterprise (i.e., small, medium, big on size and geographical scales),[6] his response vis-a-vis the operational environment (i.e., the social, economic, political context) (Alford, 1977). Each of these dimensions is an independent issue and has its parallel significance (Hoselitz, 1951 ; Timberg, 1969, 1978) and yet most of the studies tend to confuse them (Hutchins, 1958, p. 642 ; Redlich, 1957, pp. 57-63).

As a phenomenon, entrepreneurship has been analysed in the Weberian and neo-Weberian framework in western countries. Since the Indian educational system in general, and its research effort in particular, follows what becomes the focus of attention in the western countries—albeit at a later stage on the time continuum, researchers on entrepreneurship in India are largely no different in nature (and to some extent even in content) than those in the west.[7] Even later efforts are either a continuation of Weberian or neo-Weberian sociological methodology or an Indian scenario of a wider effort originating in the west.[8] In nearly all these attempts entrepreneurship is seen only in a linear fashion, with the following consequences :

1. Generalisations are offered for the entire community, linguistic, religious, ethnic, regional groups, or for the entire nation on too meagre empirical evidence. Even these scanty empirical data are not properly and/or statistically validated. Some phenomena are rejected out of hand if the researcher concerned, diving in his tiny empirical pool, surfaces even with a solitary fish of a different colour. The more enterprising among these proceed to link up their findings from Indian material, across totally alien cultures.

2. Commencement of research investigations often starts and proceeds with prior theoretical bias, thus prejudging the empirical facts. Small wonder then that most of them emerge proving what they have fervently believed all along and venture out to measure with a phenomenon slanted methodological gadgetry. Javillonar and Peters (1973, p. 317) cite the application of N.Ach. theory to Indian society as one such instance, confirming the individualistic bias of western researchers.

3. A third consequence is to take for granted any theory if

some celebrated name happens to be its originating source. This tendency of course is rooted in Indian history and possibly has some thing to do with the slavery that has been the Indians' lot for so long a period. Britto (1973, pp. 1-11), for example, indicates, how certain western values, beliefs and attitude on management (such as master destiny, independence, rationality, objectivity, participation in decision making, innovativeness, etc.) have clouded Indian thinking, and pleads for 'growing our own timber' in the field of successful indigenous managerial practices.

4. Lastly, the dominant tradition in the studies on entrepreneurship treats it as a process of challenge and response. In this process, the entrepreneur acts as a central dynamic agent seeking to achieve some combination of profit maximization and growth of his enterprise. The precise degree of entrepreneurship depends upon responsiveness to wider social and cultural conditions. In this approach factors such as markets and technology are seen in a sense as being given. It is an assumption fostered by the fact that these elements might well be common to all firms in an industry and/or beyond individual control. In evaluating the choices open to the entrepreneur, this is rarely extended to include an estimation of the relative contribution of entrepreneurship, as against other factors, to an enterprise's performance. Thus, a great many of studies on entrepreneurship are concerned with entrepreneurs *per se* and not with his business performance (Alford, 1977, pp. 116-33). Similarly, in analysing the entrepreneurial strategy and structure, the factors precipitating a crisis in an enterprise's development are not given due attention ; the response part is. In perceiving the need for and the nature of strategic decisions, an entrepreneur—particularly those in the large-scale manufacturing sector—in societies such as India act not singly but often in concert with quite a few members of his family and relatives (Cohen, 1973 ; Hazari, 1966). In the research analyses, however, the contribution of the latter is ignored.

ENTREPRENEURSHIP AS A PROCESS

From the preceding analysis it is clear that two patterns are visible in the description of entrepreneurship : one emphasises its structure and the other its functioning. While it is recognized that differences do exist between the meanings assigned to the concept of entrepreneurship by the followers of these two schools, it is also evident at the same time that in operational terms entrepreneurship involves strategic decision making which has both structural and functional dimensions. It is operated in a milieu, over a period of time, in various stages and is thus a historical process. This entrepreneurial process is a constellation of forces. The contribution of various factors at various stages of entrepreneurial continuum varies, and the process itself is not always linear (Tripathi, 1976a).[9]

Managements of large industrial enterprises are not concerned with all the decisions concerning their enterprises. They devote their attention and energies only to the basic ones among them. It is here that a classification of decisions proves helpful in distinguishing the entrepreneurial decisions from routine and ordinary decisions. A strategic decision, following Ansoff (1968, pp. 18-19) and Chandler (1962, pp. 8-12), is concerned with the relationship of the firm with its environment and maintenance of the health of the enterprise in the long run. Schumpeter (1939, p. 74) calls the carrying out of new combinations as 'enterprise' and those who carry them out as the 'entrepreneurs'. He, however, visualizes the actions of the entrepreneur as disturbing an existing equilibrium situation. He sees the entrepreneur to be one initiating change and generating new opportunities. On the other hand Kirzner (1973, p. 73) views the entrepreneur as one who initiates equilibrating changes ; bringing in to mutual adjustment those discordant elements which result from prior market ignorance. To him (1973, p. 81) the essential feature of entrepreneurship is not so much the ability to break away from the routine as the ability to perceive new opportunities that others have not yet noticed.

The categorization of functions by Ansoff, Chandler, Schumpeter and Kirzner makes it possible to identify the entrepreneurial task and eventually the entrepreneur himself. Thus, it emerges that the entrepreneurial task consists of defining the directions and amount of growth, the profitability targets, and the role of the enterprise in the environment. An individual who is the locus and source of these strategic decisions is, following Bruce (1976, p. 7), an entrepreneur. So long as this task structure remains constant it is inconsequential whether the term used is 'entrepreneur' or 'management'. Both the terms will be used interchangeably in this study keeping the task structure, as defined above, constant.

A crucial dimension in the operation of entrepreneurship as a historical process is the sanction that the management as an institution receives for its economic decisions, as strategies of combating turbulence pertaining to its enterprise.[10] In its legal form it is present in every society and certain characteristics are even found to be universally applicable. It is, however, the socio-cultural sanction that is culture or country specific. It varies in different societies. In societies such as India with which this study is concerned, this sanction is a more complex since there is an overlap of socio-cultural norms with the legal requirements in the management of enterprises. Five Indian features stand out and emphasize the need to study entrepreneurship as a collective, family phenomenon.

1. Despite depiction by many scholars, of acute centralization of authority in the hands of the *karta* in the Indian family, the fact remains that in reality strategic decision making is made by a collective authority resembling a council, and headed by the *karta*. While it is true that the *karta* does indeed play an important part in the operationalization of this authority structure, he is by no means the only locus of strategic decisions. In fact in its functioning and because of the inbuilt nature of mutual consultation, strategic decision making in the Indian industrial families becomes somewhat more diffused than is generally believed. It is, therefore, not factually correct to attribute the locus of all strategic decisions to the *karta* alone when it is known that in reality a collective body undertakes such exercises for and on behalf of the family.

2. Lewis (1955, pp. 148-74) indicates that it is the extended family that is the basic unit for individual identification and orientation. Similarly, Javillonar and Peters (1973, pp. 314-18) consider family to be the basic unit in India which determines and around which revolves most of the individual's activities throughout his life. The individual is expected to subordinate his personal wishes to the interests of the family. They argue that Indian society is a family-oriented society for which it is more meaningful to view entrepreneurship as a family rather than as an individual phenomenon.

3. The overlap of certain socio-cultural practices and personal laws with the secular laws dealing with the regulation of industrial enterprises makes it more profitable to keep the management of industrial enterprises within the family. For example, under the existing Hindu Law which recognizes partial as well as complete partition of joint family property, the family holds a privileged position with respect to Income and Wealth Taxes as compared to an individual tax payer (Gulati and Gulati, 1962, p. VII). In so far as this arrangement helps large-scale tax avoidance this privilege alone or in combination with certain other secular laws has been a great motivator in keeping the management of enterprises in India within the legally flexible ambit of the family.[11]

4. Historically, in the Indian industrial field some British institutions have helped to maintain the dominance of a few families. Among these the erstwhile managing agency system was the foremost (Chowdhry, 1966, pp. 132-47). As an institution it served the British well in the colonial days and still does so in the post-Independence period. What is more important, it has helped to legitimize the hold of one family on one enterprise. Its role in the subsequent period has strengthened the role of industrial empire-building by a few families. As a consequence much of the India's modern industry is in the hands of a few families whose operations extend to several areas (Lamb, 1956, p. 3). In spite of the existence of laws regulating licensing of new units and the expansion and operation of capacities of old ones, MRTP etc. the hold of a few families on the industrial economy continues.[12] Thus, in India the family is more than just a social institution. Within its confines it encompasses a variety of activities (social, economic, political, etc.) and organizations controlled by its members.

5. Analysing entrepreneurship as a family phenomenon also meets one of the requirements considered essential by the Marxists school in organizational sociology. This is that the family—more so an Indian family—represents a unit of class theory.

Even Schumpeter (1955, p. 113) argues that the family, not the physical person, is the true unit of class theory. Among others, Zeitlin (1974, pp. 1108-09) states that if we are to locate the actual centres of corporate control, we must discover 'the most effective kinship unit'.[13] Without research into the web of kinship relations binding apparently unrelated individuals into a cohesive unit for purposes of control, analysis of the locus of control of a large corporation is hampered at the outset. Analysis of family as a dynamic resource allocating system would enable the researcher to discern whether or not a capitalist class exists.

Because of the foregoing factors, the present study investigates entrepreneurship as a family phenomenon.

NOTES

1. The review itself is not country-specific although the studies dealing with entrepreneurship in India are given special attention.
2. Major instances of such descriptions where the ideal characteristics are located in the 'modernized' west and its bipolar one found present in the 'traditional' countries are : Toennies' (1955) 'Gemeinschaft' and 'Gasellachaft' Durkheim's (1946) 'mechanical solidarity' and 'organic solidarity' : Becker's (1957) 'Sacred and Secular' ; Redfield's (1955) 'folk' and 'urban' ; Parsons (1951) 'pattern variables' ; Lerner's (1964) 'traditional' and 'modern' ; Kluckhohan's (1955) 'man subjugated by nature' and 'man over nature' ; Merton's (1961) 'local' and 'cosmopolitan' ; Reisman's (1961) 'tradition-directed' and 'other directed' and Rostow's (1961) 'traditional' and 'mass-consumption societies'.
3. Rao and Verma (1970, pp. 67-76) in fact argue that the historical minority group status has, in a majority of cases in India, contributed to the decline of their fortune and got them reduced to the pathetic state of weaker sections.
4. McClleland himself seems to have realised this. See, for example, McClleland (1976).
5. See, for example, Tripathi (1971 and 1976).
6. See, for instance, Chowdhuri (1975) and Tyabji (1977) for an insightful discussion on the nature of enterprises.
7. How this tendency has affected research effort in India in two research areas (i.e. 'role taking' and 'action research') and what can be its consequences at the national level are shown by Verma (1961, 1969, 1974).
8. Two studies in the Weberian tradition are Singer (1972) and Ramu (1973). Among the Indian studies on management effectiveness and management styles which have their roots in the west are : England, et al. (1975), Khandwala, et al. (in progress), Nagandhi and Prasad (1971). The study of Khandwala, et al is continuation of Khandwala's earlier study conducted in Canada (Khandwala : 1976).
9. In Tripathi's schema these include : 'Social' (family, caste, comm-

unity, language, religion), 'Personal' (education, value-orientation, contacts, competence); 'Economic Environment and Infrastructure' (credit, transportation, competition, technology, demonstration effect), and 'Political-legal framework' (industrial policies of the government, national goals).

10. This, as Peterson and Berger (1971, pp. 97-107) and Zeitlin (1974, pp. 1073-1119) indicate, is vis-a-vis employees, dealers, suppliers, customers, competitors, government and the society. If the operations cross the boundaries of the nation, as they invariably do in case of large families, it would also include the same set in other nations.
11. One particular group of laws pertinent here is the one dealing with family and private trusts.
12. See, for instance, Khurana (1979).
13. Zeitlin's conceptualization of control is similar to what has been considered the management or entrepreneurial task in this study. To him (1974, pp. 1091-92) control means realization of corporate objectives over time.

3

FAMILY AND MANAGEMENT OF ENTERPRISES
Conceptual Framework

Two clusters of variables are central to the analysis attempted in this study : family, and management of enterprises in India. Of these two, undoubtedly family has received the largest attention of the researchers, Indian and foreign. This should have consequently ensured conceptual clarity and greater coverage of empirical details. What one encounters instead is a great deal of confusion about the concept of family, its typologies, and its relationship with larger societal processes such as industrialization, urbanization and development.[1] Interest in the study of the second major cluster of variables, that is, management of enterprises is of recent origin. To that extent, ambiguity about its meaning and its relationships with other variables, dependent and independent, and absence of a substantial number of studies is understandable.

This chapter will first review the studies dealing with the relationship of family with management of enterprises. The conceptual framework adopted by this study will then be outlined. Finally, the specific objectives and hypotheses proposed to be empirically tested will be specified.

STUDIES DEALING WITH THE RELATIONSHIPS OF FAMILY WITH MANAGEMENT OF ENTERPRISES IN INDIA

Before attempting a review of the studies dealing with the relationship of family with management of enterprises, it is essential to demarcate its scope. Definition of its scope is partly 'positive', indicating precisely what it would deal with ; and partly 'negative', stating what would be kept out of its purview. The following considerations have determined the scope of the review attempted here :

1. In this study, family and management of enterprises have been considered as the interacting variables. As such, the review too is concerned with only those studies which have analysed the relationship between the two.

2. The unit of analysis is not one enterprise, registered as a private or joint stock company. It is an industrial 'family'. It is useful to make a distinction between what Hazari (1966, p. 5) calls an industrial group and what is here labelled as industrial family. An industrial group is defined by Hazari as one consisting of units that are subject to the decision-making power of a common authority. This common authority in Hazari's definition not only includes members from the same family but also others, including relatives. An industrial family on the other hand, is defined here as one consisting of units that are subject to the decision-making power of a common authority that includes only members of the same family. It is recognized that the units standing in the name and under control of relatives and some others might have links—quite close ones in fact—with the industrial family and might indeed be working, at the strategic decision-making levels, in a concerted fashion. However, they are considered here as a separate industrial family for the purpose of analysis.

3. Non-empirical, impressionistic, journalistic and biographical studies are not considered. It is not that they do not offer information on the areas of concern for this enquiry. It is lack of any proper methodology and, in some cases, lack of objectivity in presenting or not presenting facts that pose problems.

4. Among the empirical studies only those that directly deal with family and management of industrial enterprises in India are to be examined. This results in the elimination of the following types of studies from the review :

(a) studies dealing with the same theme but conducted in some other countries[2] ;

(b) studies dealing with the family and management of

commercial but not industrial enterprises in India[3] ; and
(c) studies that deal with other themes but in which family business or its management is discussed in a casual manner among a whole lot of other variables.

Because of the above mentioned considerations, this review is limited to such studies which directly analyse the interesting relationship of the family and management of industrial enterprises. Only three studies are reported. These are the ones by Gaikwad and Tripathy (1970), Singer (1972) and Cohen (1973). These three studies are analysed in terms of their objectives, approach to the problem, area, scope of enquiry, hypotheses propounded if any, conclusions drawn, and observed limitations.

OBJECTIVES

The study of Gaikwad and Tripathy (1970, p. 1) was carried out with a view to examining the pattern of industrial growth in Tanuka taluka of the West Godavari district ; identifying the factors that played a decisive role in moulding industrial entrepreneurship, and suggesting steps for the effective utilization of identified factors for industrial growth in rural areas. Singer's (1972, p. 282) study was undertaken to show how a small group of successful industrial leaders in Madras city adapted their social institutions, ritual structure and belief systems to their economic activities, and how they adapted their economic activities to their cultural traditions. The purpose of Cohen's (1973, pp. 21-24) study was to analyse the problems of simultaneous membership in two different and conflicting social systems (the family and industrial units), each with its own values, norms and expectations ; the conditions under which role conflicts appeared, their behavioural consequences and the mechanisms to cope with them; their impact on managerial performance ; how the family system combined with the business system in a wider culture of change to produce emerging managerial behaviour ; how family's unity and continuity was maintained in the context of problems faced because of conflicting demands of family and business systems, when family and business systems made each other dysfunctional. In deciding the objectives of their study, Gaikwad and Tripathy did not start with any preconceived notions. Singer and Cohen, however, did so. Whereas Singer was very much a participant in the earlier on-going debate on the Weberian thesis and was, through his Madras study, out to prove empirically how wrong Weber was, Cohen was, with his self-admitted western bias (1973, p. 24), expecting to find severe inter-role conflict (1973, p. 37).

THEORETICAL APPROACH TO THE CENTRAL THEMES OF ENQUIRIES

Singer belongs to the 'neo-Weberian' tradition. Explaining his approach to the Central theme of his Madras study, Singer (1972, pp. 285-86) criticizes scholars who try to explain individual behaviour by reference to social and cultural systems of which the individual is a member. In his opinion a more fruitful approach is to begin with the ways in which particular individuals and groups make use of their family, caste, religious affiliation and networks in particular fields of activity (such as industry) and then consider what these facts imply about congruence or incompatibility of "structural types". In effect, Singer ends up in analysing the 'reciprocal effects' of urban industry on such characteristic Indian institutions as joint family, the caste system and Hindu system, and showing forms of change and persistence that transcend the dilemma faced by the "structuralists".

Gaikwad and Tripathy do not specifically state their theoretical position but a perusal of their study makes it clear that they have followed systems approach identifying socio-psychological factors that have played key roles in the entrepreneurship displayed in the Tanuku region.

Cohen (1973, pp. 26-27) has followed the inter-role conflict theory conceptualizing behaviour in terms of social systems. Central to the analysis undertaken are a set of interdependent elements, interactions, activities and sentiments, which can be seen as an organized whole. Traditionalness is predicted to be the discriminating variable in producing inter-role conflict.

AREA AND SCOPE OF ENQUIRY

The study of Gaikwad and Tripathy (1970, p. 1) is essentially a microscopic depth study that concentrates on the large-scale enterprises of one entrepreneur. It also examines a number of other entrepreneurs. Specific areas of enquiry are: personal profiles of the entrepreneur and his socio-economic background ; external factors ; nature of the enterprise, its organization and growth ; and methods used by the Department of Industries, Government of Andhra Pradesh for the promotion of industries in rural areas.

Singer's (1972, p. 290) study is wider in its scope since it covers the relationship of industrialization with the joint family, caste system and Hindu rituals and beliefs. His study is part of his wider thesis explaining "modernization" of the great Hindu tradition. In this analysis, industrialization is examined in terms of the operation of large-scale manufacturing enterprises. Singer's analysis covers aspects of family structure of successful industrial families and probes patterns of inter-generational persistence or change in them, relating them to one another and to the associated processes of urbanization, industrialization and modernization.

Cohen's (1973, pp. 42-46) study covers a 'sufficiently diverse sample' selected on the basis of community affiliation, size of family, complexity of company's size, technology and environment, and variables which affected traditionalness of family and business expectations. The selected respondents are from both the large- and small-scale manufacturing sectors and the items covered are background, environment, division of work, stated goals and objectives of the organization, locus and quality of decision making, authority relationships, family goals and objectives, family decision making, norms and values, consequences of family dynamics on the growth of the family and business, and management of conflict.

METHODOLOGY EMPLOYED

Two of the three studies are region- indeed sub-region-specific case studies. It so happened that they were conducted in South India. The locale of Gaikwad and Tripathy's study is a prosperous rural taluk, Tanuku in West Godavari district of Andhra Pradesh. Singer's study is concerned with industries in and around metropolitan Madras. Cohen's study has its eight respondent families drawn from all parts of the country.

A variety of data collection methods were used in each study. Gaikwad and Tripathy (1970, p. 3) used unstructured interviews as the main tool for getting the life histories. Personal information was obtained through a schedule, and genealogical method was used to obtain family details. General information about the industry was obtained through a structured schedule. Singer (1972, pp. 281-82) collected statistical information about the companies through official records and historical information about the families and their practices through personal interviews with the industrialists and secondary sources. Cohen (1973, pp. 42-43) opted for what he called a 'middle range of techniques'. He used open-ended questionnaires and obtained information through interviews which were sometimes guided.

All the three studies had small samples. The sample in Gaikwad and Tripathy's study consisted of 11 industrial and 15 business entrepreneurs. In the case of Singer's study there were only 19 subjects out of which two were themselves not industrialists. Cohen's study had only eight cases. The composition of the samples was revealing. Singer included only successful industrialists, determining the selection of his respondents on the basis of both official statistics and the opinion of knowledgeable persons. Gaikwad and Tripathy did not specify whether the selection of the sample was based on success and/or failure. Companies represented in Cohen's sample, as stated earlier, belonged to various communities and had specific characteristics. Singer's respondents were all large-scale industrialists. In the case of Gaikwad and Tripathy only one entrepreneur could be considered a large-scale manufacturer. Cohen's respondents represented both the large-scale and small manufacturing sectors.

HYPOTHESIS AND CONCLUSIONS

Gaikwad and Tripathy (1970, pp. 1-2) did not postulate any hypothesis as such for validation but simply stated that physical, socio-economic and political factors determined the nature of entrepreneurship in a given region. In Tanuku, they (1970, pp. 8-9) identified four types of factors which had significantly, logically and sequentially contributed to the growth of industrial and business entrepreneurship : historical and socio-economic factors, socio-biological factors, biological-legal-economic factors, and socio-psychological factors.

Singer (1972, pp. 291-350) too did not start with any specific hypothesis. His major conclusions were that : there has been a trend toward smaller nuclear families over three generations but also active cultivation of joint family contacts with relatives and maintenance of joint households in the cities, and flexible adaptation in joint family rules for industrial organization and management. The adaptation showed itself primarily in what Singer calls 'compartmentalization', 'vicarious ritualization', 'ritual neutralization of work sphere', 'secularization' and 'redefinition' of the essentials of Hinduism, and has been used by the industrialists to avoid conflict.

Cohen (1973, p. 41) did not formulate any hypothesis for empirical validation from his data but expected the characteristics of Indian families to be different from the characteristics expressly desired for business enterprises with many possible difficulties if traditional family behaviour was carried into the family business where family members desired different behaviour from themselves and others. After ranking the eight cases on the traditionalness matrix and categorizing the companies on that basis, Cohen formulated a set of 10 propositions which he thought could serve as hypothesis for investigations in other settings.

OBSERVED LIMITATIONS

The following are the observed limitations of these studies :

1. As stated earlier, the studies analysed how the socio-cultural variables impinge upon work organizations and how the interaction of the two affects each other. Further, they also attempt to relate the interactional encounters with the performance of the sample families. Measurement of performance is confined only to the industrial field. In fact, in the case of Cohen it is reduced to only one unit. However, performance in the industrial field is only a segment of the total sweep of economic activities in which the families are engaged. Crucial to the issue in particular are the development and deployment of familial resources which equally influence industrial performance. These factors are ignored by these studies.

2. The size of sample in each case is very small. Therefore, there is very little scope for generalizations at the national level. Gaikwad and Tripathy and Cohen very rightly do not attempt them. Singer, however, clearly overstates his facts when

he rebuts Weberian postulations. He might well be proved right when all-India data are assembled but on the basis of his own material, the finality of tone in his assertions and conclusions can hardly be buttressed by the empirical base of his study.

3. The sample is always designed to measure precisely the issues on hand. In this respect Gaikwad and Tripathy started with a clean slate since they did not have to prove anything. Singer's study is clearly faulty in its sampling design.[4] It is difficult to consider it as a representative sample. The same is true in the case of Cohen, whose 'all-India' sample of eight companies represented neither the industrial communities nor the complexities of the technologies used. That is perhaps one reason why he did not attempt any generalization and concentrated on working hypothesis which were to be empirically tested by others through future research.

4. Entrepreneurship being a process it is better amenable to historical analysis such as putting facts in a sequence and objectively examining the contributions made by various factors in the absence or emergence of operation of entrepreneurship. All three studies do present historical material but only Gaikwad and Tripathy present the sequential contribution of various factors to entrepreneurship on the historical continuum. The success or failure of entrepreneurship is more or less analysed in Singer's case without reference to different points of time. However, an industrial family might be considered to be among 'failures' at a particular point of time. If it is not totally wiped out as for example, in the case of Ranchhodlal Chhotalal of Ahmedabad, it is possible that it might recover and join the ranks of the 'successful'. It is also possible that a successful industrial family at present might end up in failure in the near future.

5. Of the three studies, only Cohen's analysis covers the internal dynamics of the family. However, since his empirical base is so weak, he could not offer any generalization. In the other two, very little attention has been paid to the internal dynamics of family in a historical perspective. Thus family, as Lanz, et al. (1973, pp. 582-83) argues, has largely been treated as a sponge, but with not much capacity to create change without outside intervention. While no social scientist would probably disagree with the proposition that there always is some change

occurring in the internal structure of the family, the nature of such change continues to remain obscure. It is of course true that major attention has been given to broad macroscopic variables in anticipation that some linkage between these and internal family change agents could be made. The inability to spell out such a linkage between broad macroscopic factors and internal variables in the family has led to significant gaps in the past in our knowledge of family structure and change. These gaps appear very significant when the family is viewed as a dynamic resource allocating system and as an agent triggering changes in the external systems that are dependent upon the allocation of familial resources.

CONCEPTUAL FRAMEWORK OF THE STUDY

The conceptual framework of this study has been outlined in the analysis that follows by showing first, how the family functions as a dynamic resource allocating system ; second, what factors contribute to structural change in the Indian industrial families ; third, how structural changes in turn affect the resource allocation exercise.

Family as a dynamic resource allocating system

The habit of modelling the institutions outside the family on the pattern of the family has been fairly widespread in many countries in the past. Although the practice itself has declined in quite a few countries, it still exists in certain others. In western societies where the family norms do not generally influence the structure and processes of non-familial institutions, the growing complexities of modern society have changed the nature of individual participation in the larger societal processes. From less complex and generalized participation it has become a more role-specific, universalistic and disciplined activity in a variety of structures. The family with its diffuse, particularistic and affective behaviour perhaps has become increasingly important in providing emotional support and encouragement (Bott, 1957 ; Farber, 1964 ; Goode, 1959, 1963).

It has been argued that modernization generates many specialized organizations which perform functions once associated with the family. However, the fact still remains that the family plays an important role in preparing its members to participate in these organizations (Litwak, 1963, pp. 290-323). At a time when the family members face multiple and sometimes conflicting demands affecting their behaviour and community and societal processes, the functioning of the family as a dynamic resource allocating system assumes supreme importance. In such a context, the allocating function of the family vitally affects external structures that are dependent on such allocations.

Nature of family resources

Familial resources have been categorized into four major groups (Gove, et al. 1974, pp. 182-83) :

1. Skills of family members which include skills that command a price in the market and can be sold in the employment market. In quite a few cases these constitute the primary source of family income. It has been argued that education makes a positive impact on such skills (Schultz, 1963 ; Becker, 1964, 1965). However, there are other skills such as leadership, empathy and foresight which enable the family to function effectively in handling internal and external turbulence. These are generally not marketable.

2. Package of time and energy available to family members, where both time and energy available to the members of the family are limited. As such the family must make allocations in such a manner as to ensure an adequate quantity of both for the discharge of internal functioning of the family as well as for the advancement of professional careers.

3. Set of social supports available to the family on a non-monetary basis—resources based on services provided by kinship, friendship, neighbourhood and other traditional collectivities. These resources can also be offered by a variety of governmental and non-governmental voluntary institutions.

4. Purchasing power, which could be ones own or commandeered. In many cases, this is restricted to the family's own income, but with certain others it could be considerably augmented by the availability of such power with assistance from relatives and friends, or through simple borrowing.

Resource allocation process in the family

The process of allocation of familial resources not only involves apportioning of resources among the potentially competing claims, but also decisions about a 'mix' of resources requiring a trade-off between various elements constituting the familial to resources. The trade-off itself is quite difficult to arrive at, for it triggers a series of disturbances in the working equilibrium of the family.

At any point in its history, the family could identify the components of the life style which it considers essential (Gove, et al. 1972, p. 184). Together these components could be viewed as comprising pressure points. Plotted throughout the family development cycle, a pressure line emerges. The shape of this pressure line would be entirely dependent upon the requirements or demands by the family of goods and services at different points of time. The slope of the pressure line, its height and its apex are primarily determined by internal family dynamics and external social forces. The two sets of factors have varying degrees of influence on families belonging to the different socio-economic strata.

Family income profile and pressure lines

The income profile of the family and its pressure lines both have same type of curves, and yet they do not synchronize with each other throughout the family development cycle (Gove, et al. 1972, p. 184). This discrepancy in their synchronization results in what Wilensky (1963, pp. 105-24) has very aptly described as 'life-cycle squeeze'. The squeeze itself is an eventuality that can be anticipated in advance. Efforts can be made to mitigate rather than accept it as an unavoidable catastrophe. Those with an eye for the future and the knack of planning, amass economic resource during the period of what Gove, et al. (1972, p. 185) call 'economic comfort' when the time energy, skills, financial resources and social supports exceed pressures for them. In general the periods of economic comfort occur when the size of the family is small and the life style of the family is not so high.

Factors contributing to structural change in the Indian industrial families

Resource allocation as a functional process does not operate in isolation anywhere in the world but more so in India and particularly among the industrial families. It is very much dovetailed with the biological processes of maintenance and replacement that are operative in the families. Social reproduction includes maintenance, replenishment and transmission of both human as well as social capital (knowledge, skills, values, beliefs, customs, etc.) and, viewed purely from a reproductive angle, is a cyclical process. The development of the family as a domestic group is a process that brings changes within the family and affects its relations with outside society. The reproduction function of the family generates a large number of critical events. These critical events include births and adoptions, deaths, marriages and divorces, adulthood reached and retirement. These events are called critical since they instantly affect the control over productive and reproductive resources of the families.

Impact of changes on resource allocation

Although the aforesaid critical events occur within the family, they affect the authority structure and operationalization of the internal and external business environments in the following manner : Among the industrial families in India, the resource allocation function is vested in an inner family council, called core council in the subsequent chapters, and headed by a *karta*. This authority structure handles strategic decision making for domestic purposes and also operationalizes the internal and external business environments. The control of the internal and external business environments by this authority structure is in fact exercised from the traditional seat of the family and not necessarily from the sites of work environments. Thus, the authority structure, which is essentially a family phenomenon, has penetrated into the business environment. In terms of its sweep this penetration is complete. As such the creation, maintenance and change of the business environment becomes totally dependent upon the nature and changes in this authority structure. Changes in the demographic structure of the families in turn quite often also change the authority structure. To the extent that the authority structures of the families change owing to internal events (birth, death, retirement, adulthood reached, etc.) the management structure of the industrial units gets affected. At family levels, the investment, income and expenditure structures, which are given a legal form to meet the regulatory requirements, change with almost every single change taking place in the demographic structure of the families since the latter alter positions of shares in the wealth, income and expenditure.

At the level of industrial units, the family members among others figure prominently on the formal managements. Here of course only certain types of critical events in the industrial families (retirement of an old man, death and adulthood reached) bring immediate changes since these affect the deployment of adult males in the industrial units.

Some of these critical events in industrial families also provide or alternately snap their links with other segments of society. Marriage for instance extends the relationship of the industrial families in the external environment. In the business field these new relations often tend to figure as members of the support systems inside the industrial units. They also appear as suppliers or dealers located in an external environment but having a key linkage with the industrial units, and extending important support functions to the industrial families.

The authority structure of the industrial families, as stated earlier, consists of an inner council. This authority structure is perhaps a far more stabilized structure than the demographic structure of the families. The composition profile of this authority structure of course also changes with certain types of changes in the demographic structure (death, retirement of members, inclusion of a new member). Its traditions and values which determine such key organizational issues as management styles, practices, strategies of negotiating turbulence in the families, internal and external business environments—indeed the entire content of entrepreneurship—do not change as often. However, it does change with a change in the *karta* or replacement of a key member of the core council. The impact is immediately visible in the entrepreneurial decisions inside the family and/or inside economic organizations controlled by the family.

In the Indian iudustrial families, resource allocation is as much an exercise in interpersonal collaboration as in conflict and turbulence. The functioning of the authority structure is, however, somewhat different from the western concepts, perceptions and practices (Chowdhry, 1966, p. 134). Here the exercise of the authority and power by the *karta* is somewhat diffused and a good deal of consultation is built into it. The *karta* has an inward social and outward legal authority (Prakash

Tandon cited in Chowdhry, 1966, p. 135). In the informal core council of the family, the *karta* acts as the head. The process of utilization of this authority is, however, such that generally it becomes an authority that guides rather than coerces ; critical yet co-operative ; and tolerant, and persuasive instead of dictating. There is also a considerable degree of what Chowdhry (1966, p. 135) calls familiarity, practice and skill in dealing with peer relationships. Nevertheless, it would be far from true to suggest that there are no conflicts in the familial decision making. There are conflicts—indeed quite acute and violent ones—which not only affect the internal functioning of the family but also their industrial units. In terms of their nature, familial turbulence is of three major kinds :

(a) general discord, concerning obedience and respect to the elders, observance of family traditions, religious practices performance of domestic chores, marital and personal relationships ;
(b) disagreement in the domestic sphere concerning share in the family wealth, property and income, deposits to and withdrawals from family pools, life styles ; and
(c) differences regarding economic activities on the use and exploitation of familial ventures on family versus personal guidelines.

Just as there are conflicts and problems in the industrial families, there also exist elaborate strategies of combating turbulence. Resort to these strategies does not necessarily solve the conflicts to the satisfaction of all concerned. It only helps to tide over the crisis which would otherwise have openly split the families and damaged their personality. Complete conflict resolution lies in the disappearance of discrepancies between individual and family values, needs and aspirations. This is something very difficult to attempt. It is certainly very difficult to achieve.

Viewed from a holistic point of view, the authority structures of the industrial families have to handle turbulence in their families and the internal and external business environments. Their performance consequently tends to depend upon their capacities to cope up with these tasks. Those which display overall higher capacities to deal with all three, produce the best all round results.

SPECIFIC OBJECTIVES OF THE STUDY

This study had the following specific objectives :

(a) to analyse the impact of changes in the internal structure of the industrial families on the management of enterprises ;
(b) to study the demographic and authority, and power structure of industrial families on a historical continuum ; and
(c) to identify, on the basis of the findings, areas of excellence in the management of the internal structure of the industrial families and their economic units which could be recommended for adoption in any professionally managed enterprise.

MAJOR HYPOTHESES FOR TESTING

In realizing the foregoing objectives, the study would empirically test the following major hypotheses :

1. Changes in the demographic structure of the family area are likely to change the authority structure of the family.

2. Changes in the demographic structure are likely to change the deployment of family members in the economic enterprises.

3. Nature of authority structure is likely to determine the presence or absence of turbulence in the family.

4. Business performance is likely to be dependent on the effectiveness of strategies employed to combat familial turbulence, and control internal and external business environments.

In testings these hypotheses the study would partly be filling the gaps in the research identified by Dayal (1973, p. 11), Dube (1973, p. 327c), (Dube, 1973, pp. 18-19) and Nambudiri (1973, p. 40).

NOTES

1. Shah (1973) provides a very detailed description of these inadequacies.
2. See, for example, Ishwaran (1976) for a study on family and industrialization in the Netherlands.
3. Among these, studies by Owens (1971), Ramu (1973), Sahai (1973) and Tangri (1978) are notable.
4. He (1972, p. 283) himself calls his study 'preliminary' and 'incomplete'.

4

METHODOLOGY

This chapter describes the methodology employed to achieve the specific objectives set out for the study. The issues covered are : (i) general problems faced in a sociological study using historical material ; (ii) research design, universe and sample ; (iii) tools and methods of data collection ; (iv) thematic scope of the enquiry ; (v) concepts used ; (vi) measurement of variables and (vii) limitations of the study.

GENERAL PROBLEMS FACED IN A SOCIOLOGICAL STUDY USING HISTORICAL MATERIAL

The conceptual framework of this study makes it a sociological study that utilizes essentially historical material for the empirical validation of the hypothesis. This poses two types of problems :

(a) problems involved in collection, verification, collation, restructuring and rearrangement of the available historical material ; and
(b) problems experienced in analysing the historical data because of continuing controversies.

Historical analysis requires the researcher to exercise his judgement in selecting his material, the sources ensuring its availability, and the credibility of both his facts and sources. Most of his data sources are secondary. They involve judgement of some one else, even at an earlier stage when they are recorded or formed in a particular manner. Quite often this necessitates acceptance of some and rejection of certain other sources and facts. In a few cases the researcher may find that none of the available sources offer him objective facts. In such an eventuality a reconstruction of the facts through non-contemporary sources (to the events under analysis) may become unavoidable. Cross-checking and verification of reconstructed material through a number of sources is a standardized solution in all such cases.

At a slightly later stage comes the question of collating, restructuring and rearranging the collected data. Two possibilities exist : first, where the original data can be rearranged after restructuring, and second, where the restructuring may involve exercise of judgement once again about the appropriateness and usefulness of the collected data for restructuring. The second alternative is one that quite often has been misused by a few researchers. In fairness to the research community at large, this restructuring exercise, as it has been actually done, should be suitably described so that any one having doubts about the genuineness of the exercise can check things for himself if he so desires.

Analysis of historical data in its final form is also problematic because of continuing controversies. Chirot (1976, pp. 232-41) refers to three major controversies : (i) the part played by ideas in history versus material and technological factors ; (ii) quantitative versus anti-quantitative positions and (iii) evolutionary and anti-evolutionary theorization. As he (1976, p. 236) emphasizes, there are no final judgements on these controversies. These problems can be resolved, since there is a strong thematic tradition of historical specificity among the historians and abstractions based on them are possible. The present study has attempted to resolve all these problems by cross checking facts, specifying data sources, and treating the exercise as a study in sociology using historical approach and not a study in history.

RESEARCH DESIGN

The hypothesized relationship between different variables is depicted in Diagram 4.1. The web of relationship visualized is as follows :

1. The demographic structure of the industrial families is decided by their degrees of jointness, size and composition. The structure keeps on changing because of internal demographic

Diagram 4.1
Research Design of the Study

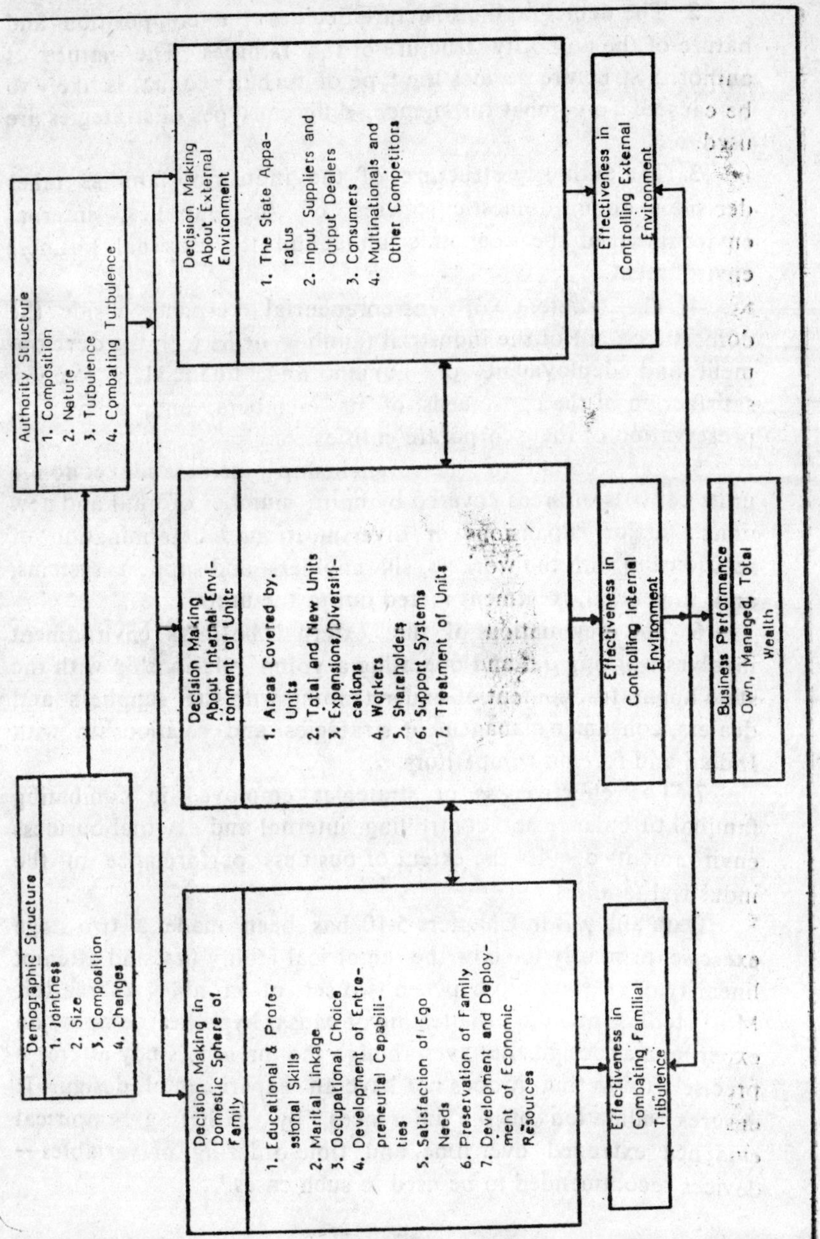

events in the families.

2. The demographic structure decides the composition and nature of the authority structure of the families. The nature of authority structure decides the type of turbulence that is likely to be caused. To combat turbulence, different types of strategies are used.

3. The authority structure of the industrial families takes decisions about domestic spheres of the families, internal environment of the economic units and the external business environment.

4. The content of entrepreneurial response inside the domestic circuit of the industrial families deals with the development and deployment of human and financial resources, satisfaction of the ego needs of its members, and, above all, preservation of their corporate entities.

5. The content of entrepreneurship inside the economic units consists of areas covered by units, number of total and new units ; major expansions or diversifications ; determination of relationship with the workers, shareholders and support systems, and, in general, treatment meted out to the units.

6. The negotiation of the external business environment involves working out and operating a going relationship with the state apparatus, content of understanding with the suppliers and dealers, consumer management strategies and relationship with Indian and foreign competitors.

7. The effectiveness of strategies employed in combating familial turbulence and controlling internal and external business environments decides the extent of business performance of the industrial families.

Data analysis in Chapters 5-10 has been made a two-stage exercise, primarily to show the empirical reality first and attempt linearity of relationship between two sets of variables thereafter. Most studies intending the testing of causal hypothesis employ an experimental design. But even though the present study attempts precisely to do that, it does not have an experimental design. It ensures validation of the inferences by gathering empirical evidence extended over time, and time-ordering of variables—devices recommended to be used in such cases.[1]

UNIVERSE AND SAMPLE

The basic unit of enquiry for this study, as stated earlier, is the industrial family. The decision itself has been made on two major considerations : first, the nature of the analysis is such as to make the family the appropriate unit of enquiry ; and second, even if it is decided to take the business house as a unit of analysis, it is very difficult to obtain certain types of data pertaining to them. Until 1969 there was, for instance, no government compiled official list of companies—private, joint stock and holding—and trusts controlled by the business houses. All previous studies dealing with industrial houses in India had relied on the information filed with the Registrar of Companies, Department of Company Affairs and the Company Affairs Division of the Reserve Bank of India. These sources, however, maintained data only for different companies. No attempt whatsoever had been made to compile information about the units controlled by various houses. The Monopoly Enquiry Commission and the Dutt Committee have done some painstaking work on this issue. Yet their reports are far from accurate and complete. After the enactment of the Monopolistic and Restrictive Trade Practices Act in 1969, a Monopolies Research Unit was established in the Department of Company Affairs, Government of India. This unit maintains a list of companies reputedly owned by various industrial houses. However, there are differences in the lists prepared by this unit and those by the Industrial Licensing Policy Inquiry Committee. Thus as of today no authoritative compilation of companies with various industrial houses is available from any source.

The problem of identifying the control of units is in itself a demanding field of enquiry. For even if the investigator succeeds in identifying the formally listed source that owns the stock of a joint stock company, there is hardly any error-proof method available to verify whether or not the formally listed source is the actual owner of the stock. Shares can be in the name of individuals, trusts, holding companies and even private and limited companies in which altogether different sets of people may have the controlling interest. The use of 'nominees' as a front for the record purposes is not only quite a common practice in India, but is practised with impunity even in western countries.[2] Thus in the formal reports submitted by registered companies, the real identity of the owners is shrewdly kept undisclosed.[3]

UNIVERSE FOR THE STUDY

The universe for this study consists of those families which meet the following criteria :

1. It is an Indian family showing complexities of jointness.
2. A minimum of three generations (including the present) must have been in the large-scale trade, business and manufacturing industry.[4]
3. The manufacturing units of the family have prominent position in at least one or more products or product groups.
4. The products of the manufacturing units of the family are either marketed all over India or in parts of the Indian Union and abroad.

The rationale behind utilization of the above criteria is to find out the contribution made by those families which have had a notable impact on industrial development in the country. In the process, a large number of industrial families not having reached that level are not included. Multinationals, their Indian subsidiaries and some large sized Indian companies too have made a significant impact—of whatever variety—on the industrialization of the country but these are not considered since the research design used for the study does not warrant their inclusion.

A list of industrial families meeting the above criteria did not exist. Therefore, a pilot study was undertaken. Information for the pilot study was collected from the Department of Company Affairs, Company Affairs Division of the Reserve Bank of India ; premier industrial financial institutions such as the Industrial Credit and Investment Corporation of India, Industrial Development Bank of India, Industrial Finance Corporation of India, Industrial Reconstruction Corporation of India, some commercial banks ; and a few agencies working in the field of advertising and marketing. A total of 96 families from various manufacturing centres of the country made the list.

SELECTION AND SIZE OF SAMPLE

The sample was selected on a random basis. The total number of respondents selected from different parts of the country comes to 33 which is 34.4 per cent of the universe.[5] This is a relatively large size of the sample. It was selected to give representation to a larger variety of industrial families meeting the same set of criteria. Five families, which were in the original sample, did not co-operate in the data collection and were substituted by similar types of families.

Of the 33 families in the final sample, 16 are from the western region, four each from the eastern, southern and central, and five from the northern region. Twenty-eight of the families are Hindus and five Jains. The Parsees and Muslims do not figure in the final sample because of their refusal to co-operate although they were in the original sample.[6]

INFORMATION SOURCES

Information about the 33 sample families was collected from the following sources :

1. *Primary Sources* : The primary sources consisted of available representatives of various generations of the same family, relatives, their friends, business colleagues, executives who have been their confidants, trade union leaders and such representatives of professional, educational, cultural and welfare institutions as had worked in close contact with them. The total number of such persons who provided information was 350.

2. *Secondary Sources* : The secondary sources were published and written material about the personal and professional lives of the representatives of sample families, available from various sources.

TOOLS AND METHODS OF DATA COLLECTION

Although the data has been collected from both primary and secondary sources, the needed information was about the same set of variables. Consequently, it had to be recorded on a similar type of instrument. Since it was realized at the very outset that information being solicited from the respondents was extensive in coverage and sensitive in nature, it would be useful to break up the data collection instrument into manageable sizes. To test this assumption, a single guide was prepared covering all the variables. The same was pre-tested on five industrial families (two from Kanpur, one each from Delhi, Ahmedabad and Bombay) not figuring in the final sample. The pre-testing exercise indicated that the guide might be better administered if it was broken down into three sections, each roughly dealing with family demographic structure, management structure and performance, and authority structure and entrepreneurship in domestic and economic spheres. Accordingly, three guides with detailed listing of the above mentioned items included under various variables were prepared and used.

Data pertaining to different sets of *kartas* were recorded in these guides, either after perusal, collation and assessment of the published or unpublished written secondary material, and/or through interviews with the respondents. A focused interview technique was used in which the discussion was limited to in depth delineation of only the covered items generally listed in the guide but quite often left largely unstructured.

Strategy employed in data collection

The following strategy was employed in data collection :

1. First, a general purpose letter with a small questionnaire seeking genealogical information about the industrial families was circulated among the respondents so as to make them aware of the nature of the enquiry and the type of help needed from them. It was anticipated that the questionnaires would not be mailed back by most of the respondents.

2. Secondary material available in the form of biographies, autobiographies, books and unpublished manuscripts and journalistic writings on the sample families was screened, collated, assessed and, to the extent applicable, was transferred to the appropriate places in the concerning guides.

3. Field work was completed between November 1977 and September 1978 and was divided into two phases. The first phase was largely exploratory in which rapport was established with at least one member of the respondent family and a general discussion about the scope of the enquiry was held with him. This helped in the identification of key resource persons who could

provide needed information about the family. Some of these resource persons were also contacted and this yielded a second set of names which proved very useful for data collection. The second phase was entirely devoted to interviewing persons identified in the first phase.

4. On many of the items about the same family, information was obtained from several sources. These were separately recorded, and afterwards the different versions were compared. Only the version corroborated by a majority of sources was accepted, since the object of data collection was ascertaining facts.[7]

The following problems were encountered in data collection :

1. Quite often the data needed were just not accessible, impossible to see and locate. They had to be reconstructed with the help of all concerned.

2. In some cases, where data did exist, they were deliberately kept 'untraceable'. Those who had access to them did not co-operate. Data had to be 'teased out' through what Goldman (1978, p. 28) called in such cases, investigative techniques.

3. One of the most difficult tasks was to obtain appointments with members of the industrial families. It was found expedient to obtain them with the help of such persons who had excellent rapport with them.

4. Almost all the respondents who co-operated in conducting the study insisted that their identity should not be disclosed even once in the data analysis. This was agreed to even though it meant sacrificing a great deal of qualitative detail.

Scope of enquiry and coverage

Thematically the main focus of this study is limited to the analysis of the interaction between the family and management of enterprises. Consequently, family and management of enterprises are two clusters of variables around which the entire analysis is interwoven. This exercise in its totality is historical. It involves presentation of interacting relationships between and among different variables at different points on the historical continuum. The beginning of the historical continuum starts from the time the sample family entered the large-scale business, trade and manufacturing industry. Its tail end covers the time when this study was conducted. The length of this historical continuum varies between a low of 56 years to a high of 110 years (Table 4.1). The average of this continuum is 84.8 years. However, it is possible to compare families with differing tenures by analysing the empirical data in terms of the tenures of various sets of *kartas*.

Three sets of *kartas* with their tenures varying between a low of 4 years to a maximum of 60 years have guided the destinies of the sample families. The averages of the tenures of the first, second and third sets of *kartas* works out to 28.7, 32.4 and 23.6 years, respectively. The third set is the present set and its tenure is reflected only up to the point at which the study was conducted. In the analysis of the empirical data, the points of time under each set of *karta* refer to the tail end. This is considered essential to enable the inclusion of all the events that have taken place during the tenures of the respective sets of *kartas*.

Since the study hopes to make a generalization about the Indian industrial families in the large-scale business, trade and manufacturing sectors, its empirical material is obtained from an all-India sample providing adequate geographical coverage. In terms of its thematic content, it makes an attempt to grasp the structure of entrepreneurial response in the internal environment of the families, their economic units, and negotiation of the external environment. To keep the sweep of the study within manageable limits, the last two areas have not been covered in too much detail.

WORKING DEFINITIONS OF CONCEPTS USED

A large number of concepts have been used in the study. Their working definitions are as follows :

1. A Family is a domestic group, the members of which share one or more of the following : (i) normative orientation, (ii) bonds of relationship, (iii) coparcenary, (iv) common authority, (v) co-residence, (vi) common hearth, (vii) common purse and (viii) participation in family-centred ceremonies. It is not a mere collection of individuals. More importantly, it is an institution having its own profile, identity and tradition, and one that performs the dynamic function of allocating and reallocating resources in the face of competing demands.

2. A Nuclear Family is one that contains one married couple

TABLE 4.1 : **Length of Stay in Large-Scale Business, Trade and Manufacturing Industry and Tenures of different sets of kartas**

Range of Stay and Tenures (in years)	Frequency and Percentages							
	Stay in Business and Industry		Tenures of *kartas*					
			Karta III		*Karta* II		*Karta* I	
	Frequency	Percentage	Frequency	Percentage	Frequency	Percentage	Frequency	Percentage
1— 10	—	—	2	6.1	—	—	—	—
11— 20	—	—	12	36.4	2	6.1	6	18.2
21— 30	—	—	12	36.4	11	33.3	12	36.4
31— 40	—	—	7	21.2	14	42.4	9	27.3
41— 50	—	—	—	—	5	15.1	5	15.1
51— 60	1	3.0	—	—	1	3.0	1	3.0
61— 70	7	21.2	—	—	—	—	—	—
71— 80	8	24.2	—	—	—	—	—	—
81— 90	6	18.2	—	—	—	—	—	—
91—100	—	—	—	—	—	—	—	—
101—110	11	33.3	—	—	—	—	—	—
Total stay of all respondent families	2797		779		1,070		948	
Average	84.8		23.6		32.4		28.7	

and their unmarried children.

3. A Joint Family is one that contains at least two couples related to each other.

4. A Lineal-Joint Family is one in which husbands are related as father and son.

5. A Collateral-Joint Family is one in which couples are related through a brother-brother tie.

6. A Lineal-Collateral Joint Family is one which contains at least three couples, related as father-son and brother-brother.

7. A Supplemental Membership of Family relations and non-relations not members of family by birth, adoption and marriage and living with the family.

8. Generational Strength refers to the number of live generations present in the family at any particular point of time.

9. Stem and Branch Households : Stem household is the seat of the family where the *karta* and the authority structure are based. Branch households are branches (of the stem household) that are housed in separate residences and have separate kitchens but share the other six characteristics mentioned earlier in the definition of family.

10. Demographic Changes refer to internal demographic events in the family such as births, adoptions, adulthoods reached, marriages, deaths and retirements.

11. Family Authority and Power Structure refers to the collective body taking all strategic decisions about the domestic and economic spheres of the family. It is headed by a *karta* and is comprised of a few senior members of the family. In terms of its nature, it is concerned with the exercise of power and authority in relation to attention and person control, allocation and reallocation of resources in domestic and economic spheres.

12. Turbulence refers to the acute problems that disturb the prevailing equilibrium in the family and in the economic spheres. Combating turbulence involves somehow getting over the consequential complications.

13. Management Structure refers to stabilized forms and systems of relationships through which the management task is being performed. These structures are to be found at the family and economic unit levels. The ones located at family level are mostly informal but functionally very important structures. These are called superstructures as they are edifices built up on the

formal and legal structures existing at unit levels. Two types of structures exist at the unit level, formal, and informal under-structures. The latter term is used because these types of structures exist below the formal structures.

14. Entrepreneurial Management Task consist of taking strategic decisions about the domestic and economic spheres of the family.

15. Entrepreneurship/Management of Enterprises is defined as a historical, non-linear process of activization of the entrepreneurial task by an entrepreneur or a collective authority structure. It is, in its essence, the content of response of the entrepreneur to various components of internal and external environment at various points of time. Because the two terms, entrepreneurship and management of enterprises, refer to the performance of management task, they are being given the same meaning in this study and would be used interchangeably.

16. Performance of Industrial Families refers to the outcome of entrepreneurial decisions and is reflected in the form of own wealth, managed wealth and total wealth of the families.

17. Structure of Exchange Relationship Between Industrial Families and Workers refers to three main aspects, (i) inadequacy in wage detemination, (ii) style of exercising dominance by the industrial families and (iii) response to the exercise of countervailing power by the workers.

18. Support Systems refer to internal support systems inside the economic units. These are categorized into functional management, finance and accounting and legal systems.

19. Amoral Familism[8] is the inability to think beyond the family.

20. Familism means according to the family and the economic organizations equal degrees of importance.

21. Fraternalism[9] involves extending family privileges to the members of economic organizations and fusing the identities of the two completely.

MEASUREMENT OF VARIABLES

Different variables have been measured as follows :

1. Jointness of Family : This includes 8 items, (i) normative orientation, (ii) bonds of relationship, (iii) coparcenary, (iv) com-

1976). In terms of their orientation, these measures relate to, (i) owners (profitability and growth), (ii) customers (quality, price, services), (iii) employees (salary, working conditions, job opportunities), (iv) state (compliance with laws and adherence to public policies) and (v) finances (solvency, liquidity). The two most notable of such scales are Karmel and Egan (1976) scale of managerial performance and Khandwala et al. (in progress) organizational effectiveness scale.

There are two major problems in using these measurement methodologies in the present study :

First, the enquiry is directly concerned with the families and their performance. Performance of industrial companies forms only a fraction—of course sizeable—of the performance of sample families. The latter also includes performance in development and deployment of familial resources from which accruals ensue ; the present study also covers them.

Second, use of the foregoing methodologies assumes that all the industrial families believe in rationality and fairplay, which amount to divorcing the empirical facts from the prevailing realities. As would be shown in the empirical chapters, a very large percentage of sample families has amply demonstrated their ability to gain even though the industrial units they control continue to fare badly.

In the ultimate analysis the real test of business performance has to be the total size of their own wealth and managed wealth. Discussions with the respondents indicated that this is exactly their viewpoint too. Performance of sample families, is, therefore, measured in terms of own, managed and total wealth expressed in crores of rupees.

10. Effectiveness of Strategies Employed to Control Internal Environment of Economic Units : It includes workers, shareholders and support systems. The three items are given a weightage of four, one and one, respectively, keeping in view their admitted importance to the sample families. Scores of 3, 2, 1 are given to high, moderate and low with a maximum possible score of 18.

11. Effectiveness of Strategies Employed to Control External Environment : It includes the state apparatus, input suppliers, output dealers, multinationals, internal competitors and consumers. The state and consumers among these have been given a

weightage of four and two, respectively, because of the importance accorded to them by respondent families. High, moderate and low ratings are given scores of 3, 2 and 1 respectively, giving a maximum posssible score of 30.

12. Effectiveness of Strategies Employed to Control Total Business Environment : Composite measure aggregating variables 10 and 11.

LIMITATIONS OF THE STUDY

The present study has the following limitations :

1. Industrial development of the country is result of contributions made by the entrepreneurs from the large, medium, small and tiny manufacturing sectors run by the private enterprise, and the same set run by the co-operative, joint and the public sectors. All these sectors and efforts are interlinked in their operational environments. Entrepreneurship is displayed in all such sectors. The present study covers only the families in the private large-scale sector.

2. In explaining the relationship of independent (family structure and authority structure) and dependent variables (management of enterprises and business performance), the present study covers only one side of this interaction, namely, how the changes in the demographic and authority structure of industrial families affect the management of enterprises. It does not attempt the analysis showing how the changes in the management of enterprises affect the industrial families.

3. Although the sweep of the study is embarrassingly wide, the analysis compresses the enormous amount of qualitative and quantitative data. Given more resources, it would have been possible to profitably expand the analysis and enrich the contents.

No single study can possibly explain every aspect of the same problem in equal detail. The same is valid for the present study. However, within its methodological framework the present study is a significant contribution in the area of sociology of organizations.

NOTES

1. See, for example, Selltiz, et al. (1965, pp. 131-33).
2. Francis (1976), Nyeman and Silberston (1978) and Zietlin (1974) provide interesting evidence of the presence of this practice in Great Britain and the United States.
3. The significance of Hazari's (1966) study lies not only in what it tells us about the inner and outer circles of the corporate private sector, but also in the fact that it could present such information at all.
4. According to the Monopolies and Restrictive Trade Practices Act, an industrial family with assets worth Rs. 20 crore is deemed a large business house. This definition has been followed in the study. However, there is a vital difference in approach in as much as the present study includes own and managed wealth in determining the size of wealth.
   This ceiling of assets has been revised in 1985 to Rs. 100 crores.
5. The families are not being identified since the respondents insisted on this stimulation.
6. Absence of Muslims and Parsees from the final sample does not pose any methodological problems in drawing the inferences from the empirical material, since religion is not one of the independent variables used in the analysis.
7. This is an accepted method in such studies which are aimed at ascertaining facts of past and present behaviour. See, for instance, Selltiz, et al. (1965, pp. 244-46, 252-53).
8. The concept was first used by Banfield (1958).
9. The term was introduced by Wallace (1974).

5

DEMOGRAPHIC STRUCTURE

In this chapter, the demographic structure of the industrial families will be discussed under the following headings: (i) characteristics indicating categories of jointness, (ii) size, (iii) composition and (iv) demographic changes.

JOINTNESS

From an extensive search of research literature on the Indian family, the following characteristics arc identified as indicators of jointness of the family, (i) normative orientation, (ii) bonds of relationship, (iii) coparcenary, (iv) common authority, (v) co-residence, (vi) common hearth, (vii) common purse and (viii) participation in family-centred ceremonies.

If one were to measure the degree of jointness of the families on an aggregated index, different scores would indicate the degree of jointness. At one end of the continuum would be the characteristics considered necessary to keep the family together. The other end would generally be the ideal stage where each of the eight characteristics mentioned above is present. Empirically, however, what is crucial here to decide is what precisely are the necessary conditions for the continued existence of a family. Some social scientists have pointed out the existence of families without a coparcenary, co-residence, common authority, common hearth, common purse and participation in the family-centred ceremonies.[1] They are also agreed that normative orientation (the feeling of belongingness) and bonds of relationship (deciding mutual dependence) constitute the absolute minimum number of characteristics.[2] In their absence a family can hardly be pronounced joint beyond any doubt.

Data from the present study shows the following patterns (Table 5.1) :

TABLE 5.1 : **Jointness in the Industrial Families**

Characteristics of Jointness	Frequency and Percentage of Families					
	Karta III		*Karta* II		*Karta* I	
	Frequency	Percentage	Frequency	Percentage	Frequency	Percentage
Single Characteristics						
1. Normative orientation	33	100	33	100	33	100
2. Bonds of relationship	33	100	33	100	33	100
3. Coparcenary	33	100	33	100	33	100
4. Common authority	33	100	33	100	33	100
5. Co-residence	8	24.2	8	24.2	33	100
6. Common hearth	8	24.2	8	24.2	33	100
7. Common purse	33	100	33	100	33	100
8. Participation in family-centred ceremonies	26	78.8	33	100	33	100
Aggregated Joints						
1. All eight	8	24.2	8	24.2	33	100
2. Six (1,2,3,4,7,8)	18	54.5	25	75.7	—	—
3. Five (1,2,3,4,7)	7	21.2	—	—	—	—
Sequential Presence of Characteristics						
1. All eight	All eight		All eight		All eight	
2. Six (1,2,3,4,7,8)	Six		Six		All eight	
3. Five (1,2,3,4,7)	Five		Six		All eight	

1. Of the eight characteristics considered, only five, namely, normative orientation, bonds of relationship, coparcenary, common authority and common purse have been present throughout. During the reign of the present set of *kartas* (*karta* III) even participation in family-centred ceremonies has ceased in about 21 per cent families.

2. Common residence and common hearth are present only in about 24 per cent families during the tenures of last two sets of *kartas*.

3. The 'ideal' type of families, where all eight characteristics are present throughout, constitutes roughly 24 per cent.

4. Roughly 54 per cent and 74 per cent families have six characteristics, namely, normative orientation, bonds of relationship, coparcenary, common authority, common purse and participation in family-centred ceremonies at the time of the third and second sets of *kartas* respectively.

5. The families with a minimum number of five characteristics total roughly 21 per cent at the time of the third set of *kartas*. This minimum includes the two characteristics identified by other social scientists (i.e., normative orientation and bonds of relationship) and three additional ones, namely, coparcenary, common authority and common purse. These families are scattered in a number of residences located in more than one city. As such, although at a lower level a segment of the family based in one city occasionally meets, eats and celebrates together, it is not possible to do so at the higher level of total family.

6. Sequentially, the families displaying a minimum of five characteristics at the time of the third set of *kartas* had six characteristics at the time of the second set of *kartas* and all eight characteristics at the time of the first set of *kartas*. Similarly, the families that show six characteristics at the time of the third set of *kartas* had the same number of characteristics at the time of the second set of *kartas*. However, they had possessed all eight characteristics at the time of the first set of *kartas*. Only the 'ideal' type of families had all eight characteristics throughout.

From the above empirical facts it is evident that four variables provide two crucial cutting points for determining the degrees of jointness among the industrial families. These two cutting points decide the upper and lower degrees of jointness. Coparcenary, common authority and common purse provide the lower cutting point ; common residence provides the upper cutting point.

Undue importance should not be given to the phenomenon of jointness among the industrial families since, as Shah notes (1973, pp. 168-69), the wealthier sections of society use joint family as an instrument for getting more wealth. However, it is essential to know whether wealth alone is associated with jointness or whether some other variables such as the volume of business, demographic changes, and size of family are also associated with it. To find out these relationships, degrees of jointness in the industrial families were correlated with, (i) own wealth, (ii) managed wealth, (iii) total wealth, (iv) total number of economic enterprises, (v) number of new economic enterprises, (vi) marriages, (vii) adulthoods reached, (viii) all demographic changes, (ix) family size and (x) number of household branches. Table 5.2 presents the resultant co-efficients. It emerges that at the time of first set of *kartas*, none of the considered variables showed any relationship with jointness. During the subsequent period, degrees of jointness of the industrial families is found to be inversely related with all ten variables considered. However, the coefficients are statistically insignificant in the case of five variables at the time of the second set of *kartas*, and four variables at the time of the present set of *kartas*. All these variables explain the economic activities and wealth of the industrial families. Thus, it is evident that as the size of families, number of household branches, adulthoods, marriages, and all demographic changes increases, degrees of jointness of the industrial families correspondingly decreases. Since degree of jointness has shown negative correlations—although statistically insignificant—with variables explaining wealth, it is also evident that with the increase in wealth the degree of jointness decreases, though it does not

TABLE 5.2 : **Correlation Coefficients of Jointness in the Industrial Families with Independent Variables**

Independent Variables	Coefficients		
	Karta III	*Karta* II	*Karta* I
1. Family size	—.875*	—.843*	0.0
2. Number of house-hold branches	—.791*	—.745*	0.0
3. Adulthood reached	—.816*	—.587*	0.0
4. Marriages	—.683*	—.759*	0.0
5. All demographic changes	—.853*	—.769*	0.0
6. Total number of economic units	—.403**	—.260	0.0
7. Number of new units	—.294	—.397**	0.0
8. Own wealth	—.232	—.206	0.0
9. Managed wealth	—.329	—.204	0.0
10. Total wealth	—.308	—.207	0.0

* Significant at 1 per cent level of confidence.

** Significant at 5 per cent level of confidence.

mean that the families completely break up.

FAMILY SIZE

The size of the family is determined by the reproductive and survival functions controlling the secretion and shrinkage of numbers. Reproduction, in turn, is dependent upon the number of marriages, fertility span of complete conjugal units and frequency of births. It is sometimes supplemented, by adoptions. Survival is more or less controlled by the dietary pattern, disease prevention and health care. In the present study, size of family in its totality is measured using the following indicators (i) own total membership and (ii) supplemental membership.

Own total membership

The total own membership of industrial families varies between a low of 18 to a high of 89 at different points of time (Table 5.3 and Diagram 5.1). At the time of the present *kartas* (*karta* III), the range of total own membership stands between the low point of 41 and high point of 89. However, roughly three-fourths of the families have a membership of over 70, with the average of total own membership itself emerging as 70.6. This shows an increase of 28.3 per cent over the total own membership of the industrial families at the time of the second set of *kartas.*

During the reign of the second set of *kartas*, the range of total own membership varied between 26 and 80. The percentage of families with membership between 26-50, 51-70 and 71-80 was 27.3, 66.7 and 6.1, respectively. The average of total own membership came to 55.3, which marked an increase of 44 per cent over the total own membership of the industrial families at the time of the first set of *kartas*.

Total own membership varied between 18 and 55 at the time of the first set of *kartas*, 27.3 per cent families had membership between 18 and 35, with 24.2 per cent each between 36 and 40, 41 and 45 and 46 and 55, respectively. The average membership was 36.5. Between the tenures of the second and third and first and second sets of *kartas*, the registered growth in total membership was 27.6 per cent and 51.7 per cent, respectively, indicating that the net accretion to the membership of the industrial families had been very impressive. It also emphasized the simultaneous

TABLE 5.3 : **Size of the Industrial Families : Own Total Membership**

Range of Membership	Frequency and Percentage of Families					
	Karta III		*Karta* II		*Karta* I	
	Fre-quency	Per-centage	Fre-quency	Per-centage	Fre-quency	Per-centage
18-20	—	—	—	—	2	6.1
21-25	—	—	—	—	3	9.1
26-30	—	—	1	3.0	3	9.1
31-35	—	—	1	3.0	1	3.0
36-40	—	—	4	12.1	8	24.2
41-45	1	3.0	1	3.0	8	24.2
46-50	2	6.1	2	6.1	7	21.2
51-55	1	3.0	3	9.1	1	3.0
56-60	2	6.1	9	27.3	—	—
61-65	3	9.1	6	18.2	—	—
66-70	3	9.1	4	12.1	—	—
71-75	8	24.2	1	3.0	—	—
76-80	8	24.2	1	3.0	—	—
81-85	3	9.1	—	—	—	—
86-89	2	6.1	—	—	—	—
Total Membership	2332		1827		1204	
Average	70.6		55.3		36.5	

increase in the complexity of the industrial families.

Supplemental Membership

Supplemental membership in the industrial families is of two kinds, (i) relations and (ii) non-relations. Both types of supplemental members increase the task of managing the family.

In the total supplemental membership of the industrial families studied the two segments of relations and non-relations varied in their strengths. Only one family at the time of the present set of *kartas* and four families at the time of first set of *kartas* did not have any relations as their supplementary members, all others did (Table 5.4 and Diagram 5.2). The number of relations with the industrial families varied between 1 and 10. About 51 per cent and 45 per cent families have 1-5 and 6-10 relations with them, respectively, at the time of the present set of *kartas*. The average number of relations in the industrial

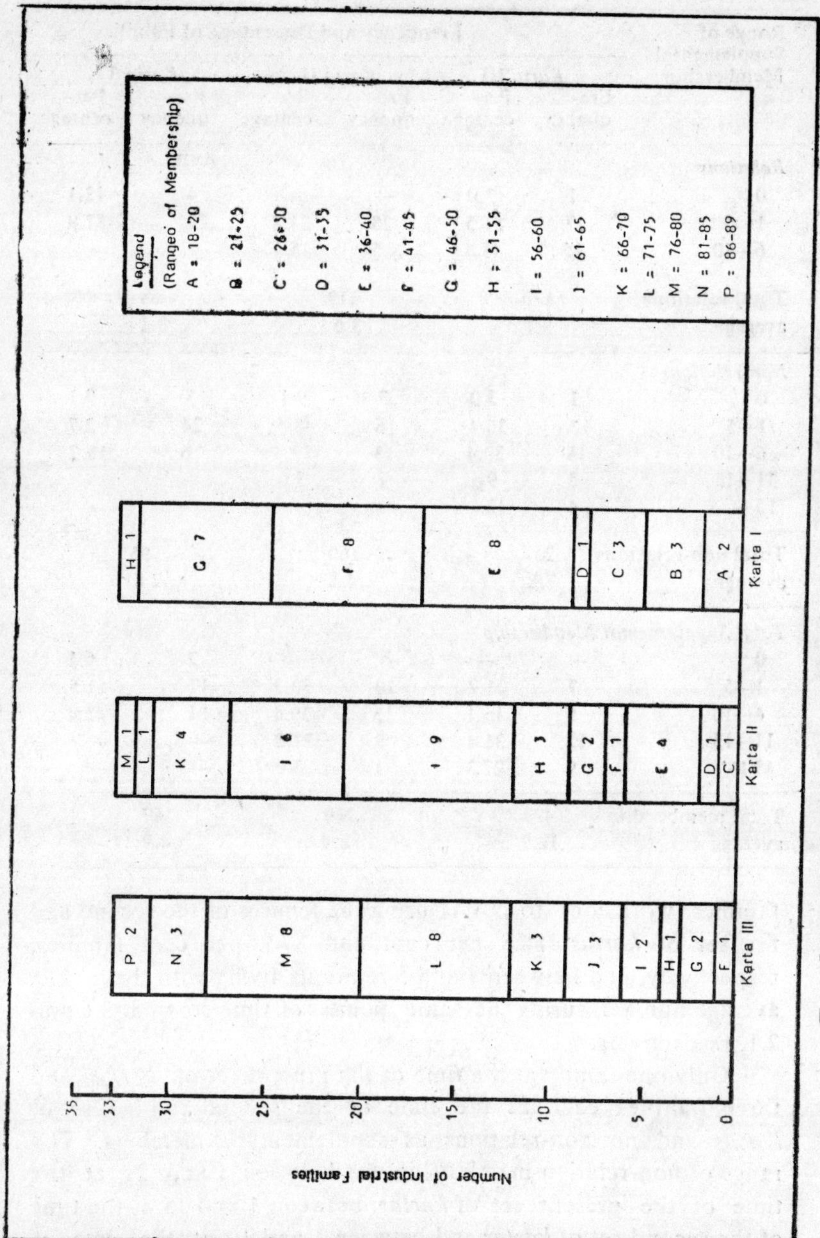

FIG. 5.1 : *Range of own total membership of Industrial Families*

TABLE 5.4 : **Supplemental Membership of Industrial Families**

Range of Supplemental Membership	Frequency and Percentage of Families					
	Karta III		*Karta* II		*Karta* I	
	Frequency	Percentage	Frequency	Percentage	Frequency	Percentage
Relations						
0	1	3.0	—	—	4	12.1
1—5	17	51.5	28	84.8	29	87.9
6—10	15	45.4	5	15.1	—	—
Total relations	170		119		68	
average	5.1		3.6		2.1	
Non-relations						
0	1	3.0	3	9.1	3	9.1
1—5	12	36.4	16	48.5	24	7 2.7
6—10	13	39.4	13	39.4	6	18.2
11—15	3	9.1	1	3.0	—	—
15+	4	12.1	—	—	—	—
Total non-relations	257		160		95	
average	7.8		4.8		2.8	
Total Supplemental Membership						
0	—	—	—	—	2	6.1
1—5	7	21.2	10	30.3	17	51.5
6—10	5	15.1	13	39.4	14	42.4
11—15	12	36.4	9	27.3	—	—
15+	9	27.3	1	3.0	—	—
Total membership	427		279		169	
average	12.9		8.4		4.9	

families works out to 5.1. During the tenures of the second and first set of *kartas*, 84.8 per cent and 87.9 per cent families, respectively, had between 1 and 5 relations living with them. The average number during the same points of time stood at 3.6 and 2.1, respectively.

Only one family at the time of the present set of *kartas* and three families each at the time of the second and first set of *kartas* had no non-relations as supplementary members. The range of non-relation membership was between 1 and 21 at the time of the present set of *kartas*, between 1 and 15 at the time of the second set of *kartas* and between 1 and 10 at the time of

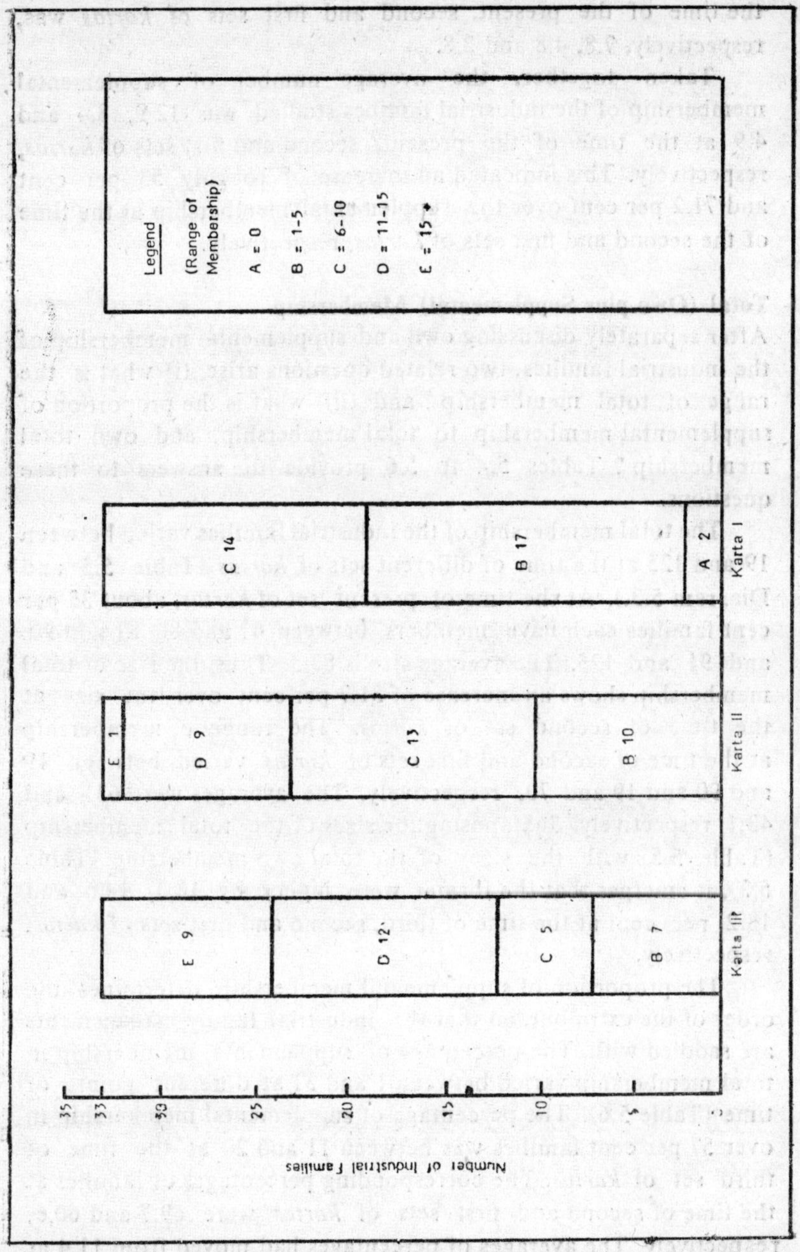

FIG. 5.2 : *Range of supplemental membership of Industrial Families*

the first set of *kartas*. The average number of non-relations at the time of the present, second and first sets of *kartas* was, respectively, 7.8, 4.8 and 2.8.

Taken together, the average number of supplemental membership of the industrial families studied was 12.9, 8.4 and 4.9 at the time of the present, second and first sets of *kartas*, respectively. This indicated an increase of roughly 53 per cent and 71.2 per cent over the supplemental membership at the time of the second and first sets of *kartas*, respectively.

Total (Own plus Supplemental) Membership

After separately discussing own and supplemental membership of the industrial families, two related questions arise, (i) what is the range of total membership ? and (ii) what is the proportion of supplemental membership to total membership, and own total membership ? Tables 5.4 to 5.6 provide the answers to these questions.

The total membership of the industrial families varied between 19 and 125 at the time of different sets of *kartas* (Table 5.5 and Diagram 5.3.). At the time of present set of *kartas*, about 33 per cent families each have members between 41 and 80, 81 and 90, and 91 and 125. The average size is 83.5. Thus, the size of total membership shows an increase of 31.7 per cent over the size at the time of second set of *kartas*. The range of membership at the time of second and first sets of *kartas* varied between 19 and 90 and 19 and 70, respectively. The averages were 63.4 and 43.1. respectively. Juxtaposing the sizes of the total membership (Table 5.5) with the sizes of the total own membership (Table 5.3), it emerges that the former were higher by 18.3, 14.6 and 18.2 per cent at the time of third, second and first sets of *kartas*, respectively.

The proportion of supplemental membership determines the order of the extra burden that the industrial family managements are saddled with. The percentage of supplemental membership in total membership varied between 1 and 31 at different points of time (Table 5.6). The percentage of supplemental membership in over 57 per cent families was between 11 and 20 at the time of third set of *kartas*. The corresponding percentages of families at the time of second and first sets of *kartas* were 69.7 and 60.6, respectively. The averages of percentages had moved from 11.4 at

TABLE 5.5 : **Total (Own plus Supplemental) Membership of Industrial Families**

Range of Membership	Frequency and Percentage of Families					
	Karta III		*Karta* II		*Karta* I	
	Frequency	Percentage	Frequency	Percentage	Frequency	Percentage
19—20	—	—	—	—	1	3.0
21—30	—	—	—	—	6	18.2
31—40	—	—	4	12.1	3	9.1
41—50	1	3.0	3	9.1	12	36.4
51—60	3	9.1	3	9.1	10	30.3
61—70	4	12.1	9	27.3	1	3.0
71—80	3	9.1	9	27.3	—	—
81—90	11	33.3	5	15.1	—	—
91—100	5	15.1	—	—	—	—
101—110	4	12.1	—	—	—	—
111—120	1	3.0	—	—	—	—
120+	1	3.0	—	—	—	—
Total membership	2758		2094		1423	
average	83.5		63.4		43.1	

the time of first set, to 13.3 at the time of second set to 14.5 at the time of present set of *kartas* indicating the gradual increase in the proportion of supplemental membership in the industrial families.

The juxtaposition of own membership with supplemental membership shows that the range of percentage of supplemental membership, as computed from the own membership, has varied between 2.4 and 54.3 at the time of present *kartas*, between 2.5 and 23.7 at the time of second set of *kartas* and between 0 and 21.2 at the time of first set of *kartas* (Table 5.7). The average of the percentages at the same points of time are 18.3, 15.4 and 12.9, respectively. Thus, it is evident that the proportion of supplemental membership has been increasing over time.

COMPOSITION OF FAMILIES

To facilitate a fuller description of the complexities of the industrial families, composition has been discussed in terms of,

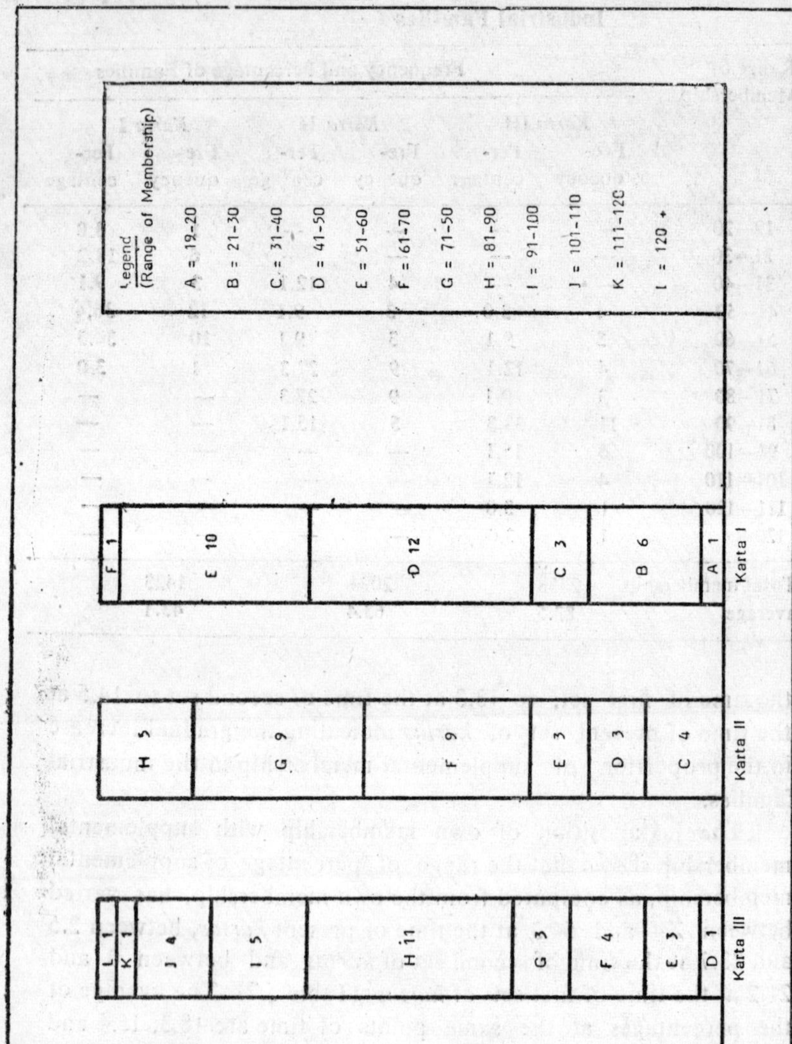

FIG. 5.3: *Range of total membership of Industrial Families*

TABLE 5.6 : **Percentage of Supplemental Membership in Total Membership of Industrial Families**

Range of supplemental membership as percentage of total membership	Frequency and Percentage of Families					
	Karta III		*Karta* II		*Karta* I	
	Frequency	Percentage	Frequency	Percentage	Frequency	Percentage
0	—	—	—	—	2	6.1
1—5	3	9.1	3	9.1	3	9.1
6—10	6	18.2	7	21.2	8	24.2
11—15	13	39.4	15	45.4	16	48.5
16—20	6	18.2	8	24.2	4	12.1
21—25	3	9.1	—	—	—	—
26—30	1	3.0	—	—	—	—
30+	1	3.0	—	—	—	—
Average	14.5		13.3		11.4	

TABLE 5.7 : **Supplemental Membership as Percentage of Own Membership**

Range of percentages of supplemental membership	Frequency and Percentage of Families					
	Karta III		*Karta* II		*Karta* I]	
	Frequency	Percentage	Frequency	Percentage	Frequency	Percentage
0	—	—	—	—	2	6.1
1—5	3	9.1	3	9.1	3	9.1
6 10	5	15.1	5	15.1	6	18.2
11—15	9	27.3	9	27.3	14	42.4
16—20	7	21.2	11	33.3	7	21.2
21—25	4	12.1	5	15.1	1	3.0
26—30	1	3.0	—	—	—	—
31—35	2	6.1	—	—	—	—
35+	2	6.1	—	—	—	—
Average	18.3		15.4		12.9	

(i) typological composition, (ii) generational strength, (iii) nature of membership (males, females and children), and (iv) stem-branch segments.

Typological Composition

Of all the possible typological categories of the families[3], the sample families displayed only the lineal joint, collateral joint, and lineal-collateral joint (Table 5.8).[4]

TABLE 5.8 : **Composition of the Industrial Families : Typological Categories**

Typological Categories* of Families	Frequency and Percentage of Families					
	Karta III		*Karta* II		*Karta* I	
	Frequency	Percentage	Frequency	Percentage	Frequency	Percentage
Lineal joint	—	—	1	3.0	2	6.1
Collateral joint	—	—	7	21.2	3	9.1
Lineal-collateral joint	33	100	25	75.7	28	84.8

* For a definition of the typological categories refer to Chapter 4.

One of the criteria of selection of an industrial family to the universe and eventually the sample used was the degrees of jointness. As such, theoretically conjugal and nuclear families have no chance of getting included in the sample. However, the important point is that even if this criterion were to be ignored, all families figuring in the universe and sample just could have not reached the stage of meeting the remaining criteria, if they were conjugal or nuclear families.

Historically, it is found that the movement of the industrial families has been from lineal to collateral to lineal-collateral joint families. At the time of present set of *kartas*, all industrial families are lineal-collateral joint. However, it does not mean that in terms of complexity all of them are alike. They are different in terms of their membership size (as seen earlier), residential norms and norms of interpersonal relations (as shall be explained later).[5]

Generational Strength

The composition of the family as analysed in terms of the generational strength at a particular point of time, shows three patterns among the industrial families (Table 5.9).

TABLE 5.9 : **Composition of the Industrial Families : Generational Strength**

Generational Strength of Families	Frequency and Percentage of Families					
	Karta III		*Karta* II		*Karta* I	
	Frequency	Percentage	Frequency	Percentage	Frequency	Percentage
Three generations	13	39.4	15	45.4	11	33.3
Four generations	17	51.5	15	45.4	22	66.7
Five generations	3	9.1	3	9.1	—	—

1. Three generations were present in over 39, 45 and 33 per cent families at the time of the present, second and first sets of *kartas*, respectively.

2. At the same points of time, four generations were present in 51.5, 45.4, 66.7 per cent families.

3. Five generations were present in about nine per cent families at the time of present and second sets of *kartas*.

Nature of Membership : Males, Females and Children.

The nature of the membership in terms of males, females and children at any given point of time provides an indication of the nature of the inter-relationships that exist in the family.[6] Plotted at different points of time it becomes evident how they have changed.

The range of male membership varied between 16 and 40 at the time of the present set of *kartas*, between 6 and 30 at the time of second set of *kartas*, and between 6 and 20 at the time of first set of *kartas*. The average number of males was 26.4, 19.2 and 13.3 during the tenures of the present, second and first sets of *kartas*, respectively (Table 5.10). The number of males at the time of present set of *kartas* was 37.5 per cent higher than the number of males at the time of second set of *kartas* ; the increase at the time of second set of *kartas* (over the number at the time of first set of *kartas*) was 44.7 per cent.

The number of females was slightly less than males throughout the historical continuum. The range of membership too was

TABLE 5.10 : **Composition of the Industrial Families : Males, Females and children**

Range of Membership	*Karta* III Frequency	*Karta* III Percentage	*Karta* II Frequency	*Karta* II Percentage	*Karta* I Frequency	*Karta* I Percentage
	Males					
6-10	—	—	2	6.1	8	24.2
11-15	—	—	5	15.1	16	48.5
16-20	5	15.1	12	36.4	9	27.3
21-25	8	24.2	13	39.4	—	—
26-30	16	48.5	1	3.0	—	—
31-35	3	9.1	—	—	—	—
36-40	1	3.0	—	—	—	—
Total males	872		634		438	
average	26.4		19.2		13.3	
	Females					
1-5	—	—	—	—	1	3.0
6-10	—	—	3	9.1	9	27.3
11-15	2	6.1	7	21.2	21	63.6
16-20	11	33.3	19	57.6	2	6.1
21-25	12	36.4	4	12.1	—	—
26-30	8	24.2	—	—	—	—
Total females	717		562		395	
average	21.7		17.0		12.0	
	Children					
1- 5	—	—	—	—	1	3.0
6-10	—	—	2	6.1	8	24.0
11-15	3	9.1	7	21.2	17	51.5
16-20	6	18.2	9	27.3	7	21.2
21-25	17	51.5	14	42.4	—	—
26-30	6	18.2	—	—	—	—
31-35	1	3.0	1	3.0	—	—
Tolal children	743		631		371	
average	22.5		19.1		11.2	

narrower : it was 1-30 at the time of the present set of *kartas* (as against 6-40 in case of males), 1-25 at the time of the second set of *kartas* (as against 6-30 in the case of males) and 1-20 at the time of the first set of *kartas* (as against 6-20 in case of males).

The female numerical deficiency was 17.8, 11.4 and 9.8 per cent at the time of present, second, and first sets of *kartas*, respectively. This was in spite of the fact that the female membership had increased by about 42.3 per cent at the time of present set of *kartas*.

The strength of the children was lower than the males by 7.8, 0.5 and 15.8 per cent at the time of the present, second and first sets of *kartas*, respectively, and was higher than the females by 37.6 per cent at the time of present set of *kartas*, and by 12.3 per cent at the time of the second set of *kartas*, but lower by 6.7 per cent at the time of the first set of *kartas*. The range of membership of the children varied between 11 and 35 at the time of present set of *kartas*, between 6 and 35 at the time of the second set of *kartas* and between 1 and 20 at the time of the first set of *kartas*. The strength of the children had increased by 70.1 per cent at the time of the second set of *kartas* over the strength at the time of the first set of *kartas*, and by 17.7 per cent at the time of the present set of *kartas* over the strength of the second set of *kartas*.

The proportion of males, females and children in the total family membership varied from time to time (Table 5.11). At the present time the percentage of males, females and children stood at 37.4, 30.8 and 31.8 in that order. The position at the time of the second set of *kartas* was : males and children 34.6 per cent each and females 30.8 per cent. During the reigns of the first set

TABLE 5.11 : **Proportion of Males, Females and Childern in Total Own Family Membership**

Types of Members	Proportion in Total Membership					
	Karta III		*Karta* II		*Karta* I	
	Frequency	Percentage	Frequency	Percentage	Frequency	Percentage
Male	872	37.4	634	34.6	438	36.4
Female	717	30.8	562	30.8	395	32.8
Children	743	31.8	631	34.6	371	30.8
Total own family members	2332	100.0	1827	100.0	1204	100.0

TABLE 5.12 : **Composition of the Industrial Families : Number and Location of Household Branches**

Range of Branches	Frequency and Percentage of Families							
	Karta III		*Karta* II		*Karta* I		Before *Karta* I	
	Frequency	Percentage	Frequency	Percentage	Frequency	Percentage	Frequency	Percentage
Number of Household Branches								
0	8	24.2	8	24.2	8	24.2	26	78.8
1- 5	8	24.2	10	30.3	15	45.4	7	21.2
6-10	11	33.3	13	39.4	10	30.3	—	—
11-14	6	18.2	2	6.1	—	—	—	—
Total branches	180		160		123		61	
average	5.4		4.8		3.8		1.8	
Location								
No. of branches	8	24.2	8	24.2	8	24.2	26	78.8
Household Branches in Single City								
1- 5	8	24.2	9	27.3	12	36.4	—	—
6-10	8	24.2	8	24.2	6	18.2	—	—
11-14	2	6.1	1	3.0	—	—	—	—
Total branches	109		104		82		26	
average	6.1		5.8		4.5		1.0	
Household Branches in More than One City								
1- 5	—	—	1	3.0	3	9.1	7	21.2
6-10	3	9.1	5	15.1	4	12.1	—	—
11-14	4	12.1	1	3.0	—	—	—	—
Total branches	71		56		41		35	
average	10.1		8.0		5.6		5.0	

of *kartas*, males, females and children constituted 36.4, 32.8 and 30.8 per cent, respectively. Thus, males always constituted the largest group with the children (present and second set of *kartas*) and females (first set of *kartas*) constituting the second largest group.

Stem and branch segments

The relationship of the household dimension of family with business and wealth, family size and demographic changes is very complex. In the present study it is considered useful first to show the number of household branches from the stems of the industrial families (Table 5.12), and indicate thereafter the relationship between the growth of household branches and a number of motivating considerations (Table 5.13).

The household dimension has two aspects : (i) location of households in cities and (ii) number of households. These two aspects are considered in combination so as to explain the empirical picture better. Following this scheme, the industrial families could be grouped under three categories :

1. Families that have remained at a single place in a single city throughout. This group is called Unilocal. About 24 per cent of the families belonged to this category (Table 5.12). It is not that they had not grown in size. Their average size was, for example, 36.2, 37.4 and 53.2 at the time of first, second and third sets of *kartas*, indicating a fair amount of growth. An interesting question in case of unilocal families was, how they had managed growing and large families at a single location. It turns out that all of them possessed large precincts to begin with. As the families grew in size, more open space was covered with buildings and family members moved into them. However, the very fact that they continued at one particular location throughout, involved a conscious strategic decision on the part of their authority structure.

2. Another group of industrial families is dispersed in several locations in single cities at the time of the present set of *kartas*. The group called Single City Multilocal constitutes 54.5 per cent of the total number of families. In the period before the tenures of the first set of *kartas*, all these families were based at one location. Spatial dispersion of branches among this group started during the tenure of the first set of *kartas*, and had continued

TABLE 5.13 : **Correlated Coefficients of Number of Household Branches in Different Types of Industrial Families with Independent Variables**

Independent Variables	Coefficients		
	Karta III	*Karta* II	*Karta* I
	Single City Multilocal (N=18)		
X^1	.184	.369	.349
X^2	.305	.205	.126
X^3	.280	.272	.229
X^4	.339	.306	—.011
X^5	.310	.558**	—.011
X^6	.543**	.051	—.372
X^7	—.145	.225	—.006
X^8	—.399	.276	—.272
X^9	.273	.394	—.200
	Plural City Multilocal (N=7)		
X^1	.651	.689	—.487
X^2	.643	.647	.507
X^3	.645	.665	—.498
X^4	.656	.713	.526
X^5	.666	.629	.526
X^6	.363	.023	.247
X^7	—.143	.509	.188
X^8	—.086	.404	.283
X^9	—.021	.479	—.182

*Significant at 1 per cent level of confidence.
**Significant at 5 per cent level of confidence.

Note : X^1=Own wealth ; X^2=Managment wealth ; X^3=Total wealth ; X^4=Total number of economic units ; X^5=Number of new units ; X^6=Marriages ; X^7=Adulthoods reached ; X^8=All demographic changes ; X^9=Family size.

thereafter. The number of branches in this group varied between 1 and 14 at the time of present and second sets of *kartas*, and between 1 and 10 at the time of the first sets of *kartas*. The average number of branches in this group was 6.1 (present *kartas*), 5.8 (second set of *kartas*). Thus, the increase in the number of branches registered at the time of the second and present sets of *kartas* (over their previous sets) was of the order of 38.1 and 5.2 per cent, respectively.

3. A third group among the industrial families is the one

that has several branches in more than one city at the time of the present set of *kartas*. This group called Plural City Multilocal constitutes 21.2 per cent of the total number of families. In the period before the first set of *kartas*, all these families were based in one city in one to five households. Spatial dispersion of family branches beyond a single city started during the tenures of the first set of *kartas*. The number of branches in this group varied between 1 and 14 (present and second set of *kartas*) and 1 and 10 (first set of *kartas*). The average number of branches stood at 10.1, 8.0 and 5.6 at the time of the present, second and first sets of *kartas*, respectively. The percentage increases registered at the time of second and present sets of *kartas* were 42.9 and 26.2, respectively. Thus, the number of branches with the plural city multilocal families was throughout higher than the single city multilocal families.

The subdivision of the sample into Unilocal, Single City Multilocal and Plural City Multilocal provided a methodological advantage in indicating the complex relationship of the number of branch households with the following variables ; (i) own wealth ; (ii) managed wealth ; (iii) total wealth ; (iv) total number of economic enterprises ; (v) number of new enterprises ; (vi) marriages ; (vii) adulthoods reached ; (viii) all demographic changes and (ix) family size. This advantage lies in the fact that the sample has a 'control' group, namely, the unilocal families that have no branches, and the other two groups that have a number of branches. Table 5.13 presents the results.

Since unilocal families have no household branches, no relationship exists between the number of households and other independent variables. However, what is interesting to note is the absence of any statistically significant relationship between the number of households and the nine independent variables among the plural city multilocal families. Only the number of new units at the time of the second set of *kartas* and marriages at the time of the present set of *kartas* are, positively and statistically, significantly related to the number of household branches. Another interesting finding is that the same independent variables have shown different relationships with the dependent variable, number of households, at different points of time, thus confirming the complexity of the phenomenon. Even after the exercise resulting in Table 5.13, it is not possible to draw any firm conclusion about the household dimension among the industrial families.

DEMOGRAPHIC CHANGES

Demographic changes in the family reflect operation of the reproduction and survival functions. In their wake, changes follow in the role-relationships within the family on the one hand and its relationship with outside society on the other.

On an average, the number of births during the tenure of the present, second and third sets of *kartas* was 12.8, 11.3 and 9.0, respectively (Table 5.14). The range of births varied between 1 and 20. Historically, the number of births has been increasing, the order of this accretion being 27.2 per cent at the time of second set of *kartas* and 12.6 per cent at the time of present set of

TABLE 5.14 : **Demographic Changes in Industrial Families**

Range of demographic changes	Frequency and Percentage of Families					
	Karta III		*Karta* II		*Karta* I	
	Frequency	Percentage	Frequency	Percentage	Frequency	Percentage
			Births			
1- 5	—	—	—	—	4	12.1
6-10	8	24.2	12	36.4	19	57.6
11-15	21	63.6	19	57.6	9	27.3
16-20	4	12.1	2	6.1	1	3.0
Total births	421		374		294	
Average	12.8		11.3		9.0	
			Deaths and Retirement			
1- 5	33	100	32	97.0	33	100
1	—	—	—	—	2	6.1
2	5	15.1	9	27.3	9	27.3
3	12	36.4	12	36.4	14	42.4
4	11	33.3	7	21.2	8	24.2
5	5	15.1	4	12.1	—	—
6-10	—	—	1	3.0	—	—
Total deaths and retirement	115		108		94	
Average	3.5		3.3		2.8	

			Adulthood Reached			
1- 5	—	—	1	3.0	6	18.2
6-10	2	6.1	14	42.4	20	60.6
11-15	6	18.2	13	39.4	7	21.2
16-20	16	48.5	5	15.1	—	—
21-25	8	24.2	—	—	—	—
26-30	1	3.0	—	—	—	—
Total adulthood reached average	607 18.4		374 11.3		282 8.5	
			Marriages (Males)			
1- 5	2	6.1	6	18.2	10	30.3
6-10	21	63.6	22	66.7	23	69.7
11-15	10	30.3	5	15.1	—	—
Total marriages (males) average	315 9.5		269 8.1		221 6.7	
			Marriages (Females)			
1- 5	2	6.1	8	24.2	15	45.4
6-10	26	78.8	22	66.7	18	54.5
11-15	5	15.1	3	9.1	—	—
Total marriages (females) average	281 8.5		237 7.2		184 5.6	
			All Marriages			
1- 5	—	—	—	—	—	—
6-10	—	—	6	18.2	11	33.3
11-15	10	30.3	10	30.3	16	48.5
16-20	15	45.4	13	39.4	6	18.2
21-25	6	18.2	4	12.1	—	—
26-30	2	6.1	—	—	—	—
Total marriages average	596 18.1		506 15.3		405 12.3	
			All Demographic Changes			
11-15	—	—	—	—	1	3.0
16-20	—	—	—	—	3	9.1
21-25	—	—	2	6.1	4	12.1
26-30	—	—	3	9.1	6	18.2
31-35	1	3.0	5	15.1	5	15.1
36-40	3	9.1	2	6.1	10	30.3
41-45	3	9.1	10	30.3	2	6.1
46-50	6	18.2	3	3.0	2	6.1

TABLE 5.14 : (*Contd.*)

51-55	6	18.2	6	18.2	—	—
56-60	8	24.2	1	9.1	—	—
61-65	3	9.1	1	3.0	—	—
66-70	2	6.1	—	—	—	—
71-75	1	3.0	—	—	—	—
All demographic changes average	1739 52.7		1362 41.3		1010 30.6	

kartas. In comparison, the range of deaths (including retirements) varied between 1 and 5 only, with the average being 3.5 (present *kartas*), 3.3 (second set of *kartas*) and 2.8 (first set of *kartas*). The increase in the death rate (including retirement) has been marginal.

The range of adulthood reached during the tenures of present set of *kartas* was between 1 and 30 ; it was between 1 and 20 at the time of second set of *kartas* and between 1 and 15 at the time of first set of *kartas*. The averages, in the same order, were 18.4, 11.3 and 8.3. The number of adulthood reached had thus registered a rise of 32.6 per cent at the time of second set of *kartas*, 32.6 per cent at the time of second set of *kartas* and 62.3 per cent at the time of present set of *kartas*.

The frequency range of marriages of both males and females, taken separately, varied between 1 and 15 at the time of the present and second set of *kartas* and between 1 and 10 at the time of first set of *kartas*. The averages for males were : 9.5 (present *kartas*), 8.1 (second set of *kartas*) and 6.7 (first set of *kartas*). The figures for females were : 8.5 (present set of *kartas*), 7.2 (second set of *kartas*) and 5.6 (first set of *kartas*). Thus, frequency of male marriages was slightly higher than female marriage throughout. Taken together, the frequency range of all marriages was 18.1, 15.3 and 12.3 during the reigns of the present, second and first sets of *kartas*, respectively. The increase registered in the frequency of all marriages over the same occurrences during the tenures of previous sets of *kartas* was 24.9 per cent at the time of the second set of *kartas* and 17.8 per cent at the time of the present set of *kartas*.

Frequency of all demographic changes varied between 11 and 75 (present *kartas*), 11 and 65 (second set of *kartas*) and 11 and 50 (first set of *kartas*). Averages of all demographic changes were 52.7, 41.3 and 30.6 during the tenures of the present, second and first sets of *kartas*, respectively. The occurrence of demographic changes had thus increased by about 34.9 per cent at the time of second set of *kartas* and 27.7 per cent at the time of present set of *kartas*.

An important question in regard to demographic changes is the relative proportion of different types of changes in all demographic changes. It appears that the proportion of different types of changes had varied at different points of time among the industrial families (Table 5.15 and Diagram 5.4). The composition at the time of present set of *kartas* is : adulthoods reached (35 per cent), marriages (33.9 per cent), births (24.3 per cent) and deaths and retirements (6.7 per cent). This was somewhat different at the time of the second and third sets of *kartas*. The order at the time of the second set of *kartas* was : marriages (37 per cent), births and adulthoods reached (27.5 per cent), and deaths and retirement (8 per cent). At the time of the first set of *kartas* the sequence was : marriages (37.5 per cent), births (27.5 per cent), adulthoods reached (26.4 per cent) and deaths and retirement (8.6 per cent).

TABLE 5.15 : **Proportion of Single Demographic Changes in All Demographic Changes**

Single Demographic Changes	Proportion in All Demographic Changes* (%)		
	Karta III	*Karta* II	*Karta* I
1. Births	24.3	27.5	27.5
2. Deaths and retirement	6.7	8.0	8.6
3. Adulthoods reached	35.0	27.5	26.4
4. Marriages	33.9	37.0	37.5

*Computed on the basis of Table 5.14.

SUMMARY

To sum up, as an important social formation the industrial

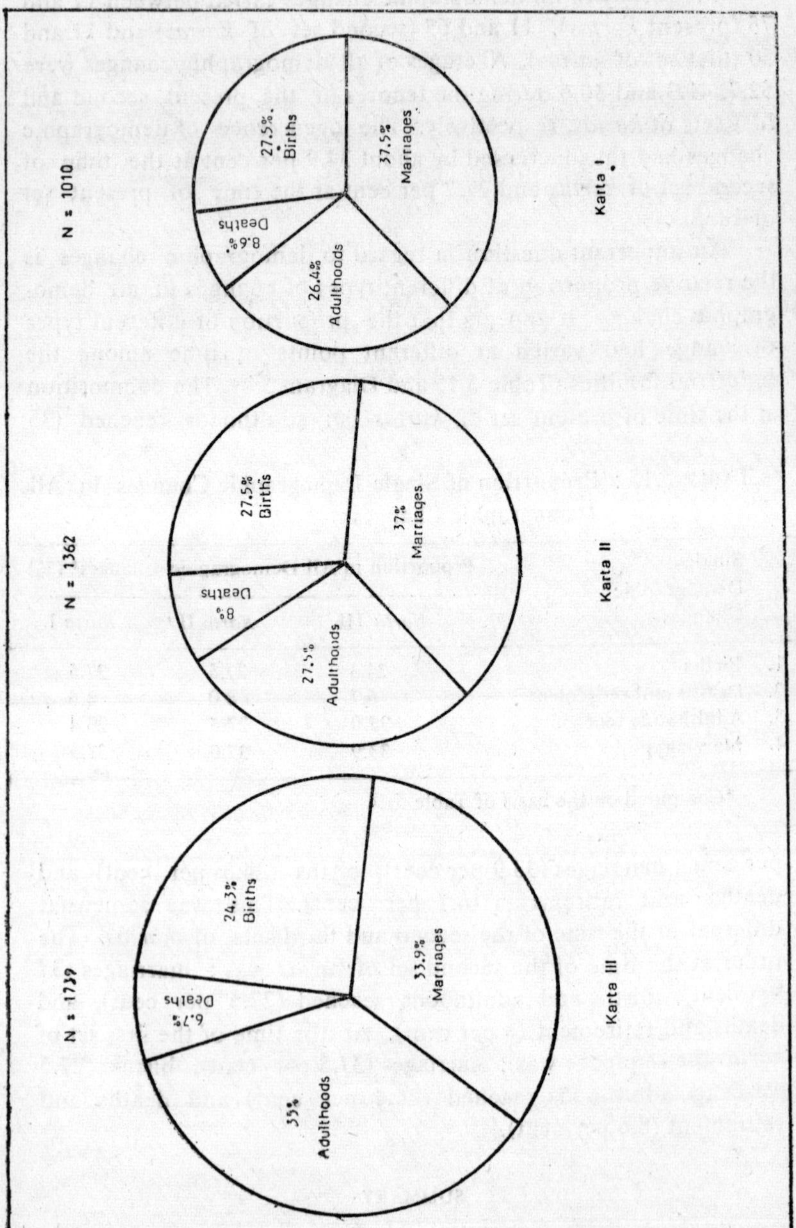

Fig. 5.4 : Proportion of different demographic changes in ... Demographic Changes

families displayed distinctiveness in a number of ways. Their degrees of jointness is crucially decided by coparcenary, common authority and common purse at its lower end and common residence at its upper level. In terms of their development process, they had shown progression :[7] their family size, own and supplemental, has continued to increase ; their complexity in terms of composition of the families in various aspects—sex, generations, typological forms—has also continued to grow, and the number of demographic changes have consistently increased.

NOTES

1. See, for example, Mukherjee (1962) and Sinha (1974, 1977).
2. *Ibid.*
3. There is no universal agreement among the social scientists about the use of compositional categories of families. Following one group of researchers, for example, the categories may be : conjungal, nuclear, lineal joint, collateral joint, and lineal-collateral joint with the possibility of these categories also being supplemental. If, however, one were to follow another stream, the family may be classified into elementary and complex. The approach used here has enabled to pursue both lines of enquiry.
4. As indicated earlier, these would have to be also deemed supplemental since an overwhelming majority among thcm has supplemental membership.
5. As Shah emphasizes (1973, pp. 86-92, 152-54), structural complexity of the family is better explained by the size of the family, residential norms and patterns of inter-personal relations.
6. An even better indicator would be exact description of the relationships (father, mother, son, daughter, grandson, daughter-in-law, etc.) as is often given in single point studies. In the present study, it was not possible to get this type of data since the study covers a large historical continuum at several points of which all these relationships have kept on changing. The only alternative left was to obtain data about composition of males, females and children.
7. The term 'progression' is used by Shah (1973, p. 82) to indicate the increase in membership of families.

6

AUTHORITY, POWER AND MANAGEMENT STRUCTURES

THE authority and power structure of the industrial family affects the behaviour of the members of the family, workers, employees of the industrial enterprises and components of the external environment. It allocates and relocates resources—human and financial—in the domestic and economic spheres of the family, and shapes family traditions. These traditions are related to child-rearing practices ; roles of males and females ; state of marital relations ; relationships among children, parents and children ; individual behaviour and adjustment ; life styles (food, clothing, housing, celebration of ceremonies, etc.), maintenance of kinship ; religious practices ; social and recreational activities, and response to public causes in the domestic field. In the economic sphere, they relate to relationships with various components of the internal and external business environment.

This chapter will cover, (i) the authority and power structure of the industrial families, (ii) the nature of turbulence in the family and mechanisms of combating them and (iii) the participation of the members of the family in the management structures of the family and its economic enterprises.

AUTHORITY AND POWER STRUCTURE OF INDUSTRIAL FAMILIES

Various studies on the nature of family authority and power structure generally cover the following three dimensions[1], (i) attention control, (ii) person control and (iii) attributed authority in the domestic and economic spheres.[2]

Since the domestic and economic spheres among the industrial families constitute two very significant areas, the third dimension was divided into two. As a result, the measure of nature of authority structure used in the following analysis, has four dimensions.

As stated earlier (Chapter 4), these dimensions have been treated as separate analytical attributes by researchers primarily because there was no significant relationship between them. In this study this premise was not accepted in advance. Whereas data were collected on separate dimensions, it was left open whether to treat them as independent attributes or as components of an aggregated scale, depending upon their empirical interrelationships. As it turned out, statistically they are, significantly correlated (Table 6.1). However, in the analysis that follows they are used both as independent attributes as well as component parts of an aggregated authority and power structure scale.

Information on these items in relation to the second and first sets of *kartas* was obtained from the present set of *kartas* and was cross-checked for accuracy with the second set of retired *kartas* (wherever available) and their friends, colleagues and confidants. In case of disagreement, only the onc obtaining maximum agreement was retained. Although response categories could possibly have been increased in a very elaborate methodological study dealing exclusively with the nature of authority and power structure, in the present study, these are limited to three analytically adequate categories of high, moderate and low. Table 6.2 presents the resultant data.

The dynamics of the nature of authority and power structure of the industrial families yielded six major patterns :

1. Families where attention and person control were moderate, and attributed authority in domestic and economic affairs high at the time of the present and second sets of *kartas*. At the time of first set of *kartas*, however, all four dimensions, namely, attention and person control, and attributed authority in domestic and economic spheres were at a high level. The percentage of the industrial families which belonged to this pattern was 21.2.

2. Families where attention and person control and attributed authority in the domestic sphere were moderate, and attributed authority in economic affairs high at the time of the present and second set of *kartas*, but where attention and

TABLE 6.1 : **Inter-Correlations between various Dimensions of Nature of Authority Structure in Industrial Families**

Variables	X^1	X^2	X^3	X^4	X^5
			Karta III		
X^1	—	.494*	.318	.182	.022
X^2		—	.643*	.418*	.881*
X^3			—	.269	.752*
X^4				—	.620*
X^5					—
			Karta II		
X^1	—	0.0	0.0	0.0	0.0
X^2		—	.644*	.418*	.885*
X^3			—	.269	.799*
X^4				—	.694*
X^5					—
			Karta I		
X^1	—	1.000*	.644*	.269	.893*
X^2		—	.644*	.269	.893*
X^3			—	.418*	.852*
X^4				—	.599*
X^5					—

*Significant at 1 per cent level of confidence.

**Significant at 5 per cent level of confidence.

Notes: X^1=Attention Control; X^2=Person Control; X^3=Attributed Authority and Power in Domestic Sphere; X^4=Attributed Authority and Power in Economic Sphere; X^5=Aggregated Authority Structure Index.

person control had been moderate and attributed authority in domestic and economic spheres high at the time of the first set of *kartas*. The percentage of the industrial families which constituted this pattern was 18.2.

3. Families where attention control and attributed authority in the domestic sphere were moderate, person control low and attributed authority in the economic sphere were at high levels during the reigns of the present and second sets of *kartas*, but where in the earlier period of the first set of *kartas*, attention and person control and attributed authority in the domestic sphere

TABLE 6.2 : **Patterns of Nature of Authority and Power Structure in the Industrial Families**

Authority Patterns	Nature* and Sequence of Authority Patterns** in Industrial Families												Frequency and Percentage of Families	
	Karta III				*Karta* II				*Karta* I				Frequency	Percentage
	A	P	D	E	A	P	D	E	A	P	D	E		
1	M	M	H	H	M	M	H	H	H	H	H	H	7	21.2
2	M	M	M	H	M	M	M	H	M	M	H	H	6	18.2
3	M	L	M	H	M	L	M	H	M	M	M	H	7	21.2
4	L	L	M	H	M	L	M	H	M	M	M	H	6	18.2
5	M	L	M	M	M	L	M	M	M	M	M	M	4	12.1
6	L	L	M	M	M	L	M	M	M	M	M	M	3	9.1

*H=High ; M=Moderate ; L=Low

**A=Attention Control ;
P=Person Control ;
D=Attributed Authority and Power in Domestic Spheres
E=Attributed Authority and Power in Economic Sphers.

had been moderate, and attributed authority in the economic sphere at high levels. About 21.2 per cent of the industrial families showed this pattern.

4. Families where, at present, attention and person control are low and attributed authority in the domestic sphere moderate and attributed authority in economic affairs high. At the time of second set of *kartas*, attention control and attributed authority in the domestic sphere were moderate, person control low and attributed authority in the economic sphere high. In the same families, attention and person control and attributed authority in the domestic sphere had been moderate, and attributed authority in the economic sphere high at the time of the first set of *kartas*. This group of industrial families totalled 18.2 per cent.

5. Families where attention control and attributed authority in domestic and economic affairs were moderate, and person control low at the time of the present and second set of *kartas*, but where all four dimensions were at moderate levels at the time of the first set of *kartas*. About 12.1 per cent of industrial families displayed this pattern.

6. Families where attention and person control are low and attributed authority in domestic and economic affairs are moderate at the time of the present set of *kartas*, but where attention control, attributed authority in domestic and economic spheres had been moderate and person control low at the time of the second set of *kartas*, and where all four dimensions were moderate at the time of the first set of *kartas*. Only 9.1 per cent of the industrial families belonged to this pattern.

In general, attributed authority in domestic and economic spheres has not changed much in the industrial families emphasizing the high degree of centralization of decision making in the allocation and relocation of resources. Only attention and person control have reached low levels. Among the possible factors responsible for the lowering of attention and person control, family size and number of household branches are considered very probable. Correlation (r) between attention and person control with these two variables was worked out to test whether indeed an inverse relationship existed between the two sets in each case. Table 6.3 presents the resultant coefficients.

It emerges that an increase in the size of family is not significantly inversely related to both attention and person control, there being no statistically significant coefficients at any point of the analysed historically continuum. Spatial dispestion of the family, as evidenced by number of household branches, on the other hand, shows a statistically significant inverse relationships throughout with both attention and person control. Thus, it is evident that lowering of attention and person control may also have been caused by the loss of constant personal contact with the family members.

TABLE 6.3 : **Correlation Coefficients between Attention and Person Control with Family Size and Number of Household Branches**

Dependent variables	Independent variables	Coefficients		
		Karta III	*Karta* II	*Karta* I
Attention control	1. Family size	—.101	0.0	—.063
	2. Number of household branches	—.497**	—.344**	—.396**
Person control	1. Family size	—.236	0.0	—.063
	2. Number of household branches	—.495*	—.524*	—.396**

*Significant at 1 per cent level of confidence.
**Significant at 5 per cent level of confidence.

As will be discussed in detail later in this chapter, the authority and power in the industrial families is vested in a collective body called the core council, headed by a *karta* and consisting of senior members of the families. It is this institution which handles strategic decisions.

NATURE OF TURBULENCE IN THE FAMILIES AND MECHANISMS USED TO COMBAT THEM

The nature of the family authority structures decides, to a large extent, the emotional well-being and satisfaction of their members, performance of roles in different areas, and pursuit of goals, values and beliefs.[3] In the ultimate analysis, these manifest themselves in the nature and extent of turbulence experienced and effectiveness of the strategies employed to combat them.

Three types of turbulence are reported among the sample families[4], (i) general familial turbulence, (ii) turbulence in the domestic sphere and (iii) turbulence in the economic sphere. On the historical continuum, three patterns of sequential occurrence of nature and extent of familial turbulence are observed (Table 6.4) :

1. Families where all three types of familial turbulence had been high throughout. Over 36 per cent of sample families belonged to this pattern.

2. Families where all three types of turbulence had been high during the reigns of first and second set of *kartas* but have become moderate at the time of present set of *kartas*. Over 36 per cent of the families belong to this pattern.

3. Families where all three types of turbulence had been high during the reign of the first set of *kartas*, but where general turbulence had become moderate during the tenures of second and third sets of *kartas*. Over a fourth of the families comprise this pattern.

The following conclusions emerge from Table 6.4.

1. The presence of turbulence in the industrial families had become almost a normal feature.

2. Turbulence in domestic and economic spheres has remained high throughout in 63.7 per cent of the families.

Discussion with the respondents indicated that moderation of general turbulence is the outcome of changes in the rules of obedience and marital traditions of the industrial families. The industrial families now have, for example, opted for a fairly complex strategy of forging links with key components of the business environment, within which viewpoints of different family members are accommodated. Possible sexual maladjustment has been moderated because the younger generation has been allowed a modicum of consultation before the selection of mates for them, and the multifarious character of their familial undertakings has afforded them the opportunities of enjoying extra-marital sex without attracting the undue attention of their spouses. The very fact that 63.7 per cent of the sample familial even now display high turbulence in domestic and economic

TABLE 6.4 : **Nature and Extent of Turbulence in the Industrial Families**

Patterns of Turbulence	Nature, Extent* and Sequence of Turbulence									Frequency and Percentage of Families	
	Karta III			*Karta* II			*Karta* I				
	General	Domestic Sphere	Economic Sphere	General	Domestic Sphere	Economic Sphere	General	Domestic Sphere	Economic Sphere	Frequency	Percentage
1	H	H	H	H	H	H	H	H	H	12	36.4
2	M	M	M	H	H	H	H	H	H	12	36.4
3	M	H	H	M	H	H	H	H	H	9	27.3

* H=High ; M=Moderate.

spheres indicates that a high price appears to have been paid by the members of the industrial families as they built up their economic enterprises.

NATURE OF STRATEGIES EMPLOYED TO COMBAT FAMILIAL TURBULENCE

Strategies employed to combat familial turbulence are decided keeping in view the nature of the turbulence, the relationship of those causing turbulence with the *kartas*, their capacity to withstand counter-measures, their susceptibilities and requirements of secrecy in handling the matter in question. In practical applications, therefore, improvizations are needed in applying basically the same strategy to different persons. It was not possible to go into the details of these improvizations, since an independent and comprehensive study would have been needed to accomplish this demanding task. Only the general nature of strategies employed and the changes that have been effected in them during the tenures of various sets of *kartas* are consequently considered appropriate for the present study.

General familial turbulence was treated with utmost firmness during the time of first set of *kartas*, since it was considered to be the touchstone of Indian industrial culture. If obedience to hierarchy was not ensured within the family itself it would have been irrational to practise it inside industrial units and in other forums of society. This was crucial in providing the authority structures sustenance in continuing the practise of always obeying the elders, irrespective of the personal beliefs of those obeying, and the nature of the actions of the former. It was also thought that if conformity were not obtained on these issues, it would overwhelm acute centralization of all decision-making powers vested in the collective authority structure. In ensuring obedience to the elders, pressure was brought to bear on the concerned persons by family members, relations and even influential members of the community. It was often accompanied by a sort of social boycott and stoppage of even normal courtesies. Although there was a fair amount of respect for the elders by the younger set in the sample families, it was also true that quite an equal amount was simply unearned. Over the years, moderation has been introduced in this strategy by reducing the role of outsiders, and generally not driving the 'offender' beyond a certain point. Quite often this means a modicum of choice being given to the individuals concerned.

Sexual maladjustment is mainly concerned with the concerned spouses, although in terms of its consequences it does quite often affect the entire family. While the practice of polygamy, keeping of concubines and extramarital sex on outstation visits were fairly widespread phenomena which were more or less tolerated during the time of the first set of *kartas*, the ones that produced serious consequences—suicides, murders, etc.—were illicit relations within the family and acute maladjustment with the spouses. In cases of polygamy, the strategy used to reduce conflict between the wives was to fix an order of precedence, duties, rights and privileges of each of the wives. In cases of illicit relations within the family, including incest, concerted efforts were made to cool off the relationship between the individuals involved. In general, no organized attempt was made to curb extramarital sex on outstation visits, although it was very vehemently denounced when detected. The situation has undergone a qualitative change after the reign of first set of *kartas*. The phenomenon of keeping several wives and concubines has almost disappeared. The incidence of in and outstation extramarital sex has, on the other hand, significantly increased. Where it has become a fairly well-known fact, the concerned family members are kept under some kind of discreet surveillance.

The strategy employed to combat turbulence concerning share in the family wealth, property and income and deposits and withdrawals from the family financial pools during the reign of first set of *kartas* was firmly to refuse giving any individual share to those causing turbulence (inheritance system under Hindu Law notwithstanding). It was possible to do so since the family income was, structurally speaking, mostly accruing not to individuals but to institutions. The entire domestic expenses came from the family pool, transfers to and from which were again authorized by the collective authority structure headed by the *kartas*. Where this strategy hurt most was in the maintenance of individualistic life styles. Although control over familial resources continued to be centralized during the reigns of the subsequent sets of *kartas*, a certain amount of liberalization in life styles, subject to certain financial ceilings, has been permitted as a strategy. Turbulence over succession has generally defied any standardized solution : all individual cases having specific contexts.

Qualitatively, turbulence in the economic sphere is altogether of a different nature from turbulence in the domestic sphere, since those causing such turbulence are not impetuous young and inexperienced members of the family. On the contrary, they are well-groomed men, who have waited in the wings long enough, suffering indifference to their wishes. These men are in possession of some control over their spheres of influence, which has certain legal props. It is certainly a very difficult task to dislodge them from these positions without causing an upheaval to the corporate entity of the family. Combating turbulence involving these types of men is a tough task that strains the nerves of the *kartas* and the collective authority structure. A popular strategy used here is to ask others in the same positions to keep a check on the dissidents to generally make it difficult for them to ignore familial norms, and not to allow them complete freedom in deciding issues. There is no practical difficulty in implementing this strategy since the boards, particularly of private and public limited companies, still maintain the pattern of familial hierarchy. The usual strategy employed, however, is direct intervention by the *kartas* or other members of the authority structure, as president/chairman of boards, and, in a few cases, by reversing their decisions and causing loss of face to them. Both these strategies continue to be used after the tenures of the first set of *kartas*.

PARTICIPATION OF FAMILY MEMBERS IN THE MANAGEMENT STRUCTURE OF THE FAMILY AND ITS ECONOMIC ENTERPRISES

Although scholars such as Drucker (1978) differentiate between management and managerial structures, quite often this distinction has not been retained by many researchers. As a result, discussions of the management structure of economic enterprises proceeds independent of the family structure of those constituting their management. It is true, of course, that management of the family and economic enterprises constitute two segments of the total management task. However, an *a priori* assumption that they constitute two separate unimpinging structures, as was done by Cohen (1973) and Singer (1972) results in a biased approach to the measurement of an empirical reality in India. In the present study, no such assumption was made and the question was left to be decided by empirical facts.

A great deal of probing and intensive discussion with the respondents, their colleagues and confidants indicated that their is no separate management structure for the allocation and relocation of resources and for combating turbulence in domestic and economic spheres among the industrial families. As a result, the content of management response to the turbulence faced in the domestic and economic spheres is the same in its abstract form, although in specificities it may be different. Since management of the domestic sphere also is largely concerned with the deployment/redeployment of familial resources of various types, it was considered as part of the entrepreneurial profile.

MANAGEMENT STRUCTURES

The participation of members of the family in various management structures depends upon the purpose for which the latter are created, that is, investment, production, etc. on the one hand, and their number on the other. There are some structures in which every family member participates, if only for record purposes. In certain others, only the adults and males do so.

The management structures operative in the industrial families are a very interesting combination of the formal and informal. In general, however, the informal structures at the top rule supreme and the formal structures help the informal structures to take decisions and implement the strategies charted out. At the bottom level, the informal understructures receive a fair degree of attention although there is a time when they get precedence over the formal structures.

The management structures of the industrial families and the enterprises managed by them comprise six hierarchies (Chart 6.1) : two among these function at the family and four at the economic unit levels. However, despite the large number of hierarchies, the distance between them is not much and the

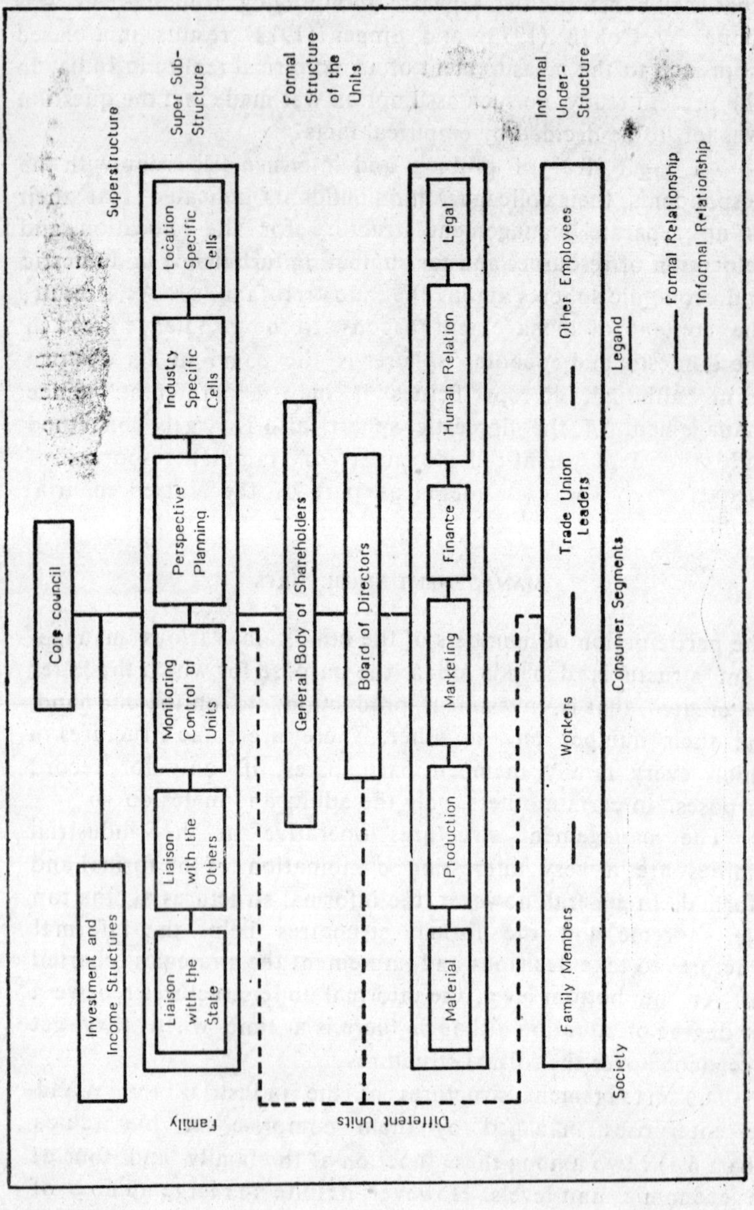

FIG. 6.1 : *Management structures in Industrial Families*

structures at the top and bottom are fairly effectively integrated both horizontally and vertically. Theoretically, the unit level managements can receive equal attention and recognition. However, as facts stand, management structures of different units have been given varying degrees of importance and attention at various points of time depending upon their capacity to yield profits : very high, good, marginal or poor. Whereas the relationship of the formal management structures at the unit levels with the superstructure working at the family level is formal and visible that of informal understructures consisting of an assorted variety of individuals including family members does not have any formal and legal existence. The operation of the former is highly regimented, the latter is quite free and flexible.

MANAGEMENT SUPERSTRUCTURE

The form, jurisdiction and recruitment management structure in the industrial families consists of two bodies : these would be henceforth referred to as the core council and the substructures because of their size and functional importance. These two structures are different from the board of directors and the general bodies of private and public limited companies in terms of the following characteristics :

1. Recruitment to the core council and substructures are based on seniority and ability to get results. Membership in both bodies is certainly not through any election.

2. Both the core councils and the substructures are not institutions for legitimization of decisions taken elsewhere. On the other hand, decisions taken at the core councils and substructures are sent to the boards and general bodies of the respective companies where the controlling capacities of the industrial families ensure their formal adoption. Thus, their jurisdiction is *original.*

3. Although the substructures are larger structures, functionally they are not mechanisms operational for reviewing, revising or approving decisions taken by the core councils.

TABLE 6.6 : **Size and composition of super-substructures at family levels**

Composition of Super Substructure	Size of Substructures (Averages)		
	Karta III	*Karta* II	*Karta* I
1. Core council members	4.3 (36.1)	4.1 (40.2)	3.4 (39.5)
2. Other family members	1.4 (11.8)	1.5 (14.7)	1.9 (22.1)
3. Managers	6.2 (52.1)	4.6 (45.1)	3.3 (38.4)
Total membership of super substructures	392	339	285
Average	11.9	10.2	8.6

Note : Figures in the parenthesis show percentages.

kartas) to 52.1 per cent (present set of *kartas*), whereas that of the core council members marginally declined (from 39.5 to 40.2 to 36.1). The percentage of family members who were not members of the core councils came down to 11.8 (present set of *kartas*) from 22.1 (first set of *kartas*).

Structures at the unit level

Formal structures : The second layer of management structures consisting of four hierarchies is at the unit level. One among these four, namely, the general bodies of the shareholders is left out of the analysis here as this is not a study of concentration, combination and interloculatory directorship.[5] The analysis about the remaining three is in terms of their composition, strength and deployment of family members on them.

The organizational form of the economic units is a combination of partnerships, private and public limited companies. During the tenures of the first set of *kartas* when all their units were established/acquired, partnerships and private limited companies constituted 51.8 and 25.5 per cent, respectively (Table 6.7). In the subsequent period the organizational form of old and new units displayed different patterns. Partnership as an organizational device is in disuse not only in the case of new units ; even the old partnerships have been converted into private/public limited companies. The percentage of old private limited companies

TABLE 6.7 : **Type and number of organizational form at unit levels**

Type of Organizational Forms	Number of Organizational Forms				
	Karta III		*Karta* II		*Karta* I
	Total	New	Total	New	Total & New
1. Partnerships	—	—	—	—	276 (51.8)
2. Private limited companies	494 (39.1)	120 (33.1)	497 (55.2)	78 (21.2)	136 (25.5)
3. Public limited companies	768 (60.9)	242 (66.9)	403 (44.8)	289 (78.8)	121 (22.7)
Total number of units	1262	362	900	367	533

Note : Figures in the parenthesis show percentages.

was higher (55.2) than public ones (44.8) during the tenures of the second set of *kartas*. Subsequently, this order was reversed. In the case of new companies at the time of both the second and third set of *kartas*, the percentage of public limited companies was 78.8 and 60.9, respectively.

Deployment of family members on the formal managements of the economic units is made in order to keep familial control over their functioning, provide training to the young entrants, and ensure protection of familial interests. The range of family members deployed on the formal managements of the economic units varies between 11 and 32 at present and varied between 6 and 30 at the time of second set of *kartas* and between 6 and 20 at the time of first set of *kartas*. The average of total deployments during the same points of time were 23.9, 17.5 and 12.5, respectively (Table 6.8). Initially the width of membership range thus was very narrow. It underwent considerable expansion at the time of second and present sets of *kartas*. The increase in the number of family members deployed on formal management of the economic units was of the order of 40.1 per cent at the time of second set of *kartas* and 36.4 per cent at the time of present set of *kartas*.

Two important issues vis-a-vis deployment of family members

TABLE 6.8 : **Number of family members deployed in economic units**

Range of Family Members Deployed	Frequency and Percentage of Families					
	Karta III		*Karta* II		*Karta* I	
	Fre-quency	Per-centage	Fre-quency	Per-centage	Fre-quency	Per-centage
6—10	—	—	2	6.1	9	27.3
11—15	2	6.1	7	21.2	19	57.6
16—20	4	12.1	19	57.6	5	15.1
21—25	12	36.4	4	12.1	—	—
26—30	13	39.4	1	3.0	—	—
30+	2	6.1	—	—	—	—
Total members deployed	790		579		412	
Average	23.9		17.5		12.5	

on the formal management of the economic units are : What is the proportion of deployed family members when compared to own total membership ? What is the proportion of the deployed family members when compared to the total number of males in the families ? Tables 6.9 and 6.10 present these proportions. The percentage of family members deployed stood between 21 and 39, with the average at the time of present, second and first sets of *kartas* being 34.5, 31.1 and 33.7 per cent, respectively. Thus,

TABLE 6.9 : **Proportion of family members deployed on the formal managements of economic units**

Range of Percentage of Family Members Deployed	Frequency and Percentage of Families					
	Karta III		*Karta* II		*Karta* I	
	Fre-quency	Per-centage	Fre-quency	Per-centage	Fre-quency	Per-centage
21—25	—	—	—	—	1	3.0
26—30	4	12.1	13	39.4	7	21.2
31—35	24	72.7	16	48.5	19	57.6
35+	5	15.1	4	12.1	6	18.2
Average		34.5		31.1		33.7

TABLE 6.10 : **Proportion of males deployed on the formal management of economic units**

Range of Percentages of Males Deployed	Frequency and Percentage of Families					
	Karta III		*Karta* II		*Karta* I	
	Frequency	Percentage	Frequency	Percentage	Frequency	Percentage
75—80	—	—	2	6.1	1	3.0
81—85	4	12.1	1	3.0	1	3.0
86—90	12	36.4	10	30.3	6	18.2
91—95	17	51.5	14	42.4	12	36.4
96—100	—	—	6	18.2	13	39.4
Average		90.4		90.1		95.6

around one third of the family membership is deployed by industrial families on the formal management. However, as Table 6.10 shows it rather vividly, this deployment is overwhelmingly male oriented.

Another crucial dimension of deployment of family members is the maintenance of family hierarchy (Table 6.11). Usually the deployment is a mix of members of core councils, substructures and ordinary members (from the families), and non-family members (from outside). Table 6.11 establishes that over 80 per cent of senior positions have always been held by the members of the core councils, and substructures, whereas ordinary family members figure mostly in comparatively junior positions. Also noteworthy is the fact that at the time of present set of *kartas*, non-family members form as much as 22.6 per cent and 31.2 per cent, respectively of the two top echelons. Thus, whereas the proportion of senior positions held by the senior members of industrial families has more or less remained stabilized around 90 per cent, with the increase in the proportion of outsiders in these positions the share of ordinary family members has somewhat declined.

Informal understructures. Maximization of profits through the pursuit of rationality, efficiency and productivity has not been the route always used by most of the industrial families. In the absence of these means, maximization of profits has become excessively dependent upon strategies aimed at the control of the various

TABLE 6.11 : **Positions occupied by family members and non-family members on the formal managements of economic units**

Type of members	Type of Positions held*											
	Karta III				*Karta* II				*Karta* I			
	A	B	C	D	A	B	C	D	A	B	C	D
A *Family Members*												
1. Core council members	352 (36.1)	414 (47.8)	752 (39.0)	210 (7.2)	324 (39.4)	340 (44.2)	615 (42.4)	180 (8.1)	248 (48.5)	252 (51.4)	369 (37.8)	120 (9.0)
2. Members of substructures	532 (54.5)	332 (38.2)	789 (40.8)	515 (17.6)	423 (51.4)	312 (40.5)	480 (33.0)	414 (18.7)	210 (41.1)	168 (34.3)	482 (49.4)	367 (27.7)
3. Ordinary members	92 (9.4)	121 (14.0)	391 (20.2)	2207 (75.2)	76 (9.2)	118 (15.3)	358 (24.6)	1620 (73.2)	53 (10.4)	70 (14.3)	125 (12.8)	840 (63.3)
Total Family members	976	867	1932	2932	823	770	1453	2214	511	490	976	1327
B. Non-family members	285 (22.6)	394 (31.2)	590 (23.4)	851 (22.5)	77 (8.6)	130 (14.4)	347 (19.3)	486 (18.0)	22 (4.1)	43 (8.1)	90 (8.4)	272 (17.0)
C. Total positions	1261	1261	2522	3783	900	900	1800	2700	533	533	1066	1599

*A=President/Chairman/Chairman-Managing Director, B=Vice President/Managing Director/Joint Managing Director ; C=Executive/Wholetime Director ; D=Ordinary Members.

Note : 1. Since one family member occupies many positions, the number of own family members (Table 6.8) and membership of core councils and substructures (Tables 6.4 and 6.5) would not tally with the deployment figures in this Table.

2. Figures in the parenthesis show percentages.

components of the internal and external business environments. Management of the internal environment is the responsibility of the formal management. However, the actualization of it is bolstered with the activization of informal understructures on the one hand, and use of the state apparatus on the other. These informal understructures generally are supplementing mechanisms, but if necessary they can well be converted into ones that can supplant the formal structure.

In pure compositional terms, these informal understructures comprise the family members, workers, other employees and trade union representatives. In their functioning, these structures are not so visible and have no legal basis. Yet they are very vital for the protection of the interests of the industrial families. In general, their numbers are larger than the actual number of the economic units, indicating the fact that one industrial unit has more than one such structure (Table 6.12). Of the four components, the trade union leadership forms the largest single block of membership, with the percentage ranging between 41.3 (present *kartas*) and 43.8 (first set of *kartas*). Other employees and workers contribute roughly around 25 per cent each (Table 6.12). The percentage of family members figuring in these understructures has declined over the years from a maximum of 8.8 at the time of first set of *kartas* to 4.3 at the time of present set of *kartas*. The prominent presence of trade union leadership in these mechanisms indicates their conflicting public and private postures on the issue of protection of workers' interests. The same is true in the case of workers and other employees who participate in these structures. Some among the latter participants are relatives or men from the same castes/communities as the industrial families. The rewards for these men include out-

TABLE 6.12 : **Number of informal understructures at unit levels**

Type of Structures	Number of Informal Understructures		
	Karta III	*Karta* II	*Karta* I
1. Total economic units	1262	900	533
2. Informal understructure	3155	1710	853
Average number of informal understructures	2.5	1.9	1.6

TABLE 6.13 : Composition and strength of informal understructures at unit levels

Composition of informal understructures	Strength of Informal Understructures					
	Karta III		*Karta* II		*Karta* I	
	Number	Average	Number	Average	Number	Average
1. Family members	329 (4.3)	10.0	321 (5.9)	9.7	288 (8.8)	8.7
2. Workers	2020 (26.6)	61.2	1294 (23.9)	39.2	766 (23.3)	23.2
3. Other employees	2109 (27.8)	63.9	1399 (25.6)	42.1	792 (24.1)	24.0
4. Trade union leaders	3138 (41.3)	95.1	2416 (44.6)	73.2	1439 (43.8)	43.6
Total membership	7596	230.2	5430	164.2	3285	99.5

Note : Figures in the parenthesis show percentages.

of-turn increments and promotions but generally no cash payments. It is an altogether different story in case of trade union representatives. The repondents indicated that there were very few trade union leaders belonging to their industrial units who did not have some understanding with them. In fact, they went to the extent of asserting that the type of life style the trade union leaders maintained was hardly possible on the salary that they were earning.

INVESTMENT STRUCTURES

While it is open to the members of the industrial families to invest in real estate, business and trade and manufacturing in their individual capacities, they have not done so simply because it would have forced them to pay penal rates of taxes. The investment structures have, therefore, been established within the ambit of existing laws (dealing with the Hindu undivided families, charitable institutions and industrial development) so that tax benefits can be obtained to the maximum extent. Their size, composition and number is quite often determined on similar considerations.

No single set of structure is popular for purposes of investment among the industrial families. On the other hand, a complex combination of investment structure has been used. A bewildering characteristic of these structures is that there is a considerable degree of overlap-plug among them. It is, for example, an accepted practice that one investment structure, also 'invested' in one or many other investment structures. During discussions it was revealed that this was a device through which the industrial families had quite often successfully avoided the identification of a particular unit as belonging to them. This enabled them to escape the regulatory measures stipulated under the Monopolistic and Restrictive Trade Practices Act. This being so, it is inevitable that the investment structures of the industrial families would also have a large number of fake nominees—*benami*—among their members. Although none of the respondents was willing to give any details on the question, the existence of this phenomenon, and its large-scale practice, nevertheless, was confirmed without much hesitation.

Much as it was desired, it was not possible to obtain information about the extent of overlapping of the investment structures among themselves since this would have revealed the strategies used by the industrial families. The data presented in Table 6.14 reflect only the formal and official factual position as it stands before the eyes of the law. Admittedly, it does not reveal the exact empirical situation. Analysing the data with this reservation, it emerges that the investment structures of industrial families broadly can be classified into two categories : private and public. The family and private trusts belong to the first category, and holding the public limited companies belong to the second. Among the first set, the difference between a family and private trust is only in terms of the demographic background of their members. While the former consists only of the family members, the latter also includes non-family members. Among the second set, the holding companies are different from the public limited companies in as much as they are originators of the public limited companies and are formed only for investment purposes. The public limited companies are established primarily for manufacturing purposes and the investments made are only incidental. Both contribute to the abstract goal of profit maximization, but in different ways.

It appeared that during the time of the first set of *kartas* a larger percentage of investment structures consisted of trusts, both family and private, among the industrial families. During the tenures of second set of *kartas*, the increase in the number of the investment structures was entirely accounted for by the holding and public limited companies, whereas the same is distributed among the family trusts at the time of third set of *kartas*. The number of the investment structures had appreciated by 14.1 per cent at the time of the third set of *kartas*, and by 41.5 per cent at the time of second set of *kartas*.

FAMILY INCOME AND EXPENDITURE STRUCTURES

Juxtaposing Tables 6.14 and 6.15, showing investment and income/expenditure structures, respectively, four important findings emerge :

1. All the family, private trusts and holding companies also are simultaneously income/expenditure structures for the industrial families.

TABLE 6.14 : **Type of Number of Investment Structures**

Type of Investment Structures	Number of Investment Structures					
	Karta III		*Karta* II		*Karta* I	
	Number	Average	Number	Average	Number	Average
1. Family trusts	73 (33.5)	2.2	46 (24.1)	1.4	46 (34.1)	1.4
2. Private trusts	46 (21.1)	1.4	46 (24.1)	1.4	46 (34.1)	1.4
3. Holding companies	46 (21.1)	1.4	46 (24.1)	1.4	30 (22.2)	0.9
4. Public limited companies	53 (24.3)	1.6	53 (27.7)	1.6	13 (9.6)	0.4
Total Number of Structures	218	6.6	191	5.8	135	4.1

Note : Figures in the parenthesis indicated percentages.

TABLE 6.15 : **Type and Number of Income/Expenditure Structures**

Type of Income/ Expenditure Structures	Number of Income/Expenditure Structures					
	Karta III		*Karta* II		*Karta* I	
	Number	Average	Number	Average	Number	Average
1. Family, private trusts and holding companies	165 (83.3)	5.0	138 (80.7)	4.2	122 (78.7)	3.7
2. Hindu undivided families	33 (16.7)	1.0	33 (19.3)	1.0	33 (21.3)	1.0
Total	198	6.0	171	5.2	155	4.7

Note : Figures in the parenthesis shows percentages.

2. All public limited companies as investment structures also are in themselves not income/expenditure structures.

3. The total mix of income/expenditure structures also includes the important legal entity known as the Hindu Undivided Family (HUF).

4. The total number of investment structures is always higher than the total number of income/expenditure structures. Thus, the number of investment structures that also are income and expenditure structures has been increasing, whereas the number of HUFs has not.

Like the investment structures, the income/expenditure structures also are legal structures. Functionally, most non-adult males and females have nothing to do with the investment and income/expenditure structures, although it does not necessarily mean that their names do not figure in the nomenclature of these structures. Distribution of family income, like the making of investment, ensured through the legal formations discussed above, is a legal exercise. Income is almost uniformly not distributed among the various members of the family as per the legal stipulations of the investment and income/expenditure structures. In effect, it means that the demographic size and complexity of the industrial families is used as a convenient device in the process of capital accumulation. While familial income is not strictly distributed on legal lines, domestic expenses for running a kitchen, housing, clothing, education, transport and entertainment have to be constantly met. This requires disbursement from a family pool to meet these expenses. The decisions about the actual details of these expenses are made by different arrangements among different types of families at different points of time. Among the unilocal families the *karta* himself decides—in consultation with one or two females—the lower and upper limits of expenses on major items. Among the multilocal families where the stem and the branch households live as separate households, the *karta* exercises the power of disbursing the money in case of the single city multilocal but its actual use gets somewhat personalized as the various branch households want. In the case of the plural city multilocal families, its members, who are also *kartas* for one location, perform the duties of the main *karta* at these locations. The stem of the family is the personal charge of the main *karta* at a place where he himself lives in both varieties of multilocal families.

To sum up, the industrial families display a high degree of centralization of authority and power in the hands of the core councils and the *kartas*. The presence of a high degree of turbulence has also become a normal feature among them. As such, a large number of strategies have been employed to combat different types of turbulence. Six hierarchies are being used for the management of the family and its economic enterprises. These hierarchies are a mix of formal and informal structures.

NOTES

1. Hadley and Jacob (1973, 1976) provide an exhaustive survey of literature on the subject.
2. A detailed description of measurement of authority structure can be seen in Chapter 4.
3. See, for example, Devereaux (1962) and Geismar (1973).
4. A detailed discussion of measurement of turbulence can be found in Chapter 4.
5. This area, and issues pertaining to it, are fairly well researched. Reference for example, may be made to the studies of Chowdhuri (1975), Industrial Licensing Policy Enquiry Committee (1969), Hazari (1966), Mahnot (1962) and Mehta (1952, 1961.)

7

NATURE OF ENTREPRENEURSHIP
Management of Familial Resources

THIS chapter examines the nature of the entrepreneurship displayed by the industrial families in managing their familial resources. This largely concerns the domestic sphere where their authority and power structures are concerned with the development and deployment of human and financial resources. The human resource structure is examined by considering :

(a) the educational attainments of the *kartas* ;
(b) the educational attainments of males and females ;
(c) the professional education/training of the males ;
(d) the family background of brides for sons and grooms for daughters ;
(e) the occupational profile before entry in the manufacturing industry ;
(f) the number of members in other occupations in the post-entry period ;
(g) the development of entrepreneurial capabilities in family members ;
(h) the satisfaction of the ego needs of family members ; and
(i) the preservation of the corporate entity of the family.

The financial resource structure is analysed by examining own, managed and total wealth on the one hand, and performance in the economic field at different points of time on the other.

EDUCATIONAL ATTAINMENTS

The educational attainments of the *kartas*, males and females present three different identities. The level of educational attainments of the *kartas* was pretty low at the time of first set when 32.4 per cent were just literates (Table 7.1). However, it was soon realized by them that if the expanding industrial empire was to be managed to their best advantage, apart from family members, they themselves would need higher education. This realization apparently found expression in the higher educational attainments of the subsequent sets of *kartas*. As much as 40.5 per cent of second set of *kartas* were graduates and over. This proportion reached its peak of 100 per cent in case of the present set of *kartas*.

In general, the unequal status of males and females is reflected in their educational attainments. A large proportion of females had lower educational attainments than the males at the time of first set of *kartas*. Educational attainments of both sexes rose in general at the time of subsequent sets of *kartas*. However, those of the males did so appreciably (Table 7.1). In spite of the general all-round improvement in the educational attainments among the industrial families, the order of importance—*kartas*, males and females—continues to show itself in the levels of educational attainments among various types of family members.

PROFESSIONAL EDUCATION/TRAINING

Managing large-scale manufacturing enterprises requires capable professional managers in various functional areas such as production, finance, marketing, personnel, etc. The data (Table 7.2) indicates that the industrial families also have felt the need for acquiring professional management capabilities within their families. At the time of first set of *kartas*, all family members deployed on the management of the economic units were only professionally acquainted with the technological processes employed there. In the subsequent period, however, a mix of both professionally acquainted and professionally qualified persons has been built up in the families in the ratio of 2 : 1. The motivation behind this strategy was revealed in the course of intimate discussions. It was indicated that the industrial

TABLE 7.1 : **Educational Attainments of Kartas, Males and Females**

Level of Educational Attainment	Frequency and Percentages of *Kartas*, Males and Females								
	Karta III			*Karta* II			*Karta* I		
	Karta	Males	Females	*Karta*	Males	Females	*Karta*	Males	Females
1. Illiterate	—	—	—	—	41 (6.5)	83 (14.8)	1 (3.0)	50 (11.4)	120 (30.4)
2. Just literate	—	143 (16.4)	122 (17.0)	—	102 (16.1)	228 (40.6)	10 (30.3)	142 (32.4)	152 (38.5)
3. School leaving certificate	—	332 (38.1)	340 (47.4)	14 (42.4)	234 (36.9)	83 (14.8)	16 (48.5)	125 (28.6)	91 (23.0)
4. Graduates and over	33 (100)	397 (45.5)	255 (35.6)	19 (57.6)	257 (40.5)	168 (29.8)	6 (18.2)	121 (27.6)	32 (8.1)
Total Number	33	872	717	33	634	562	33	438	395

Note : Figures in the parenthesis show percentages.

TABLE 7.2 : **Professional Qualifications of the Family Members Deployed on the Economic Units**

Nature of Professional Qualifications	Frequency and Percentage of Family Members					
	Karta III		*Karta* II		*Karta* I	
	Frequency	Percentage	Frequency	Percentage	Frequency	Percentage
1. Professionally acquired	535	67.7	401	69.3	412	100
2. Professional training/degree from India	125	15.8	79	13.6	—	—
3. Professional training/degree from abroad	130	16.5	99	17.1	—	—
Total number of members deployed	790	100.0	579	100.0	412	100.0

families did not consider it essential for all of their members to acquire professional qualifications to manage the industries.

FAMILY BACKGROUND OF BRIDES AND GROOMS

Apart from births and adoptions, marriages are the only other means through which recruitment to and from the families takes place, vitally affecting, in the process, their human resource structure. Additionally, marriages also serve as a device in integrating elements of business environment which keep on causing turbulence and which the industrial families, as a class try to negotiate as much as they possibly can. To the extent that marriages forge new relationships with key allies in society they strengthen the hold of the industrial families on their economic units. Data from this study provides powerful support to this postulation (Tables 7.3 and 7.4).

It has, for example, been discovered that the industrial families enter into marital alliances overwhelmingly with the following occupational segments of society, (i) businessmen, (ii) industrialists, (iii) professionals (in medicine, law, engineer-

TABLE 7.3 : **Family Background of Brides for the Sons of the Industrial Families**

Family Background of Brides	Frequency and Percentages of Brides					
	Karta III		*Karta* II		*Karta* I	
	Fre-quency	Per-centage	Fre-quency	Per-centage	Fre-quency	Per-centage
1. Businessman	32	16.6	54	20.1	72	32.6
2. Industrialists	134	42.5	81	30.1	61	27.6
3. Professionals	36	11.4	68	25.3	36	16.3
4. Bureaucrats	50	15.9	27	10.0	21	9.5
5. Politicians	29	9.2	20	7.4	—	—
6. Others	14	4.4	19	7.1	31	14.0
Total number of brides	295	100.0	269	100.0	221	100.0

TABLE 7.4 : **Family Background of Grooms for the Daughters of the Industrial Families**

Family Background of Grooms	Frequency and Percentage of Grooms					
	Karta III		*Karta* II		*Karta* I	
	Fre-quency	Per-centage	Fre-quency	Per-centage	Fre-quency	Per-centage
1. Businessman	76	27.0	44	18.6	51	27.8
2. Industrialists	96	34.2	94	39.6	65	35.3
3. Professionals	75	26.7	58	24.5	56	25.0
4. Bureaucrats	24	8.5	17	7.2	10	5.4
5. Politicians	9	3.2	6	2.5	—	—
6. Others	1	0.4	18	7.6	12	6.5
Total number of grooms	281	100.0	237	100.0	194	100.0

ing, management), (iv) bureaucrats and (v) politicians. Very few alliances are being entered into with other occupational segments of society. A second characteristic displayed is that the pattern of marital alliances of the sons and daughters of the industrial families with various types of allies significantly varies during the reigns of different sets of *kartas*. At the time of the first set of *kartas*, for example, the largest percentage of brides (32.6) for the sons were provided by businessmen as against 27.6 per cent from among industrialists. Professionals and others provided 16.3 per cent and 14 per cent, respectively (Table 7.3). In contrast, the largest percentage of groom for daughters (35.3) were provided by industrialists, with businessmen (21.8) and professionals (25) coming next (Table 7.4). Politicians did not figure in marital alliances of either sex during this period.

During the reigns of the second and third sets of *kartas* this base of class consolidation was enlarged. This enlargement involved the inclusion of politicians in the marital alliance network and somewhat more normalized distribution of the marriages among other occupational segments. This exercise itself was carried out keeping the changing relative importance at various points of time of various occupational segments to the industrial families. At the time of the second set of *kartas*, professionals, for example, provided roughly one-fourth of both brides and grooms ; bureaucrats and politicians contributed 10.0 per cent and 7.4 per cent of the brides, but only 7.2 per cent and 2.5 per cent of the grooms, respectively ; industrialists gave 30.1 per cent of the brides and 39.6 per cent of the grooms as against businessmen who gave 20.1 per cent of the brides and 18.6 per cent of the grooms.

During the tenures of the present set of *kartas*, the importance of businessmen providing brides has declined (from 20.1 per cent for the second set of *kartas* to 16.6 per cent for the present set of *kartas* and that of grooms has increased (from 18.6 per cent for the second set of *kartas* to 27.0 per cent for the present set of *kartas*). In comparison, the pattern is just the reverse in the case of industrialists who are now providing a larger percentage of brides (42.5 against 30.1) and a smaller percentage of grooms (34.2 against 39.6). Professionals are providing a lesser percentage of brides (11.4 against 25.3) but a higher percentage of grooms (26.7 against 24.5). The share of bureaucrats and politicians has appreciated in both types of alliances but this appreciation in the case of bureaucrats is more marked in providing brides rather than grooms.

During discussions with the respondents it was revealed that although marital alliances were being entered within the same caste during the reign of first set of *kartas* this restriction gradually had been removed. In fact, at the time of third set of *kartas* marriages even outside their religion have been tolerated. Whereas generally sons were not consulted in the selection of brides during the reign of first set of *kartas*. Subsequently, the practice of consulting them had begun to be adopted. However, class consolidation was and continues to be the overriding consideration in clinching marital alliances among the industrial families.

OCCUPATIONAL STRUCTURE

By being in the manufacturing sector, the industrial families were pursuing one major cluster of occupations. However, a more complete picture of their historical occupational profile can be presented by examining its structure before they entered the manufacturing industry and the nature of non-manufacturing occupations in which their members had been engaged after their entry.

That a very large percentage of present industrial families in India was engaged in business, trade, banking and money-lending during their pre-industrial existence, and entered the manufacturing industry after accumulating a large enough amount of capital is a fairly well-documented fact.[1] The data from the present study also confirm this trend (Table 7.5). In addition, it is also possible to identify the complementary role played by three different configurations of main and subsidiary occupations in the process of capital accumulation and industrial development along the capitalist path. The first of these configurations, constituting about 51.5 per cent, consisted of trade and business as main occupations, and four subsidiary occupations, namely, white and blue collar jobs[2], professions (such as law, medicine, engineering) and senior government positions which were of course common in the remaining two configurations as well. In this configuration, the more paying type of two subsidiary occupations, that is, professions and senior government service, accounted for 27.2 per cent as against 24.2 by the other two. Banking and money-lending figured as the main occupations in the second configuration, accounting for roughly 30.3 per cent sample families. In 18.2 per cent of these families, professions and senior government service emerged as subsidiary occupations.

TABLE 7.5 : Configuration of Main and Subsidiary Occupations of the Industrial Families before their Entry into Manufacturing Industry

Configuration Pattern	Frequency and Percentage of Families						
	Main Occupations			Subsidiary Occupations			
	Business and Trade	Banking and Moneylending	Landlord and Farming	White Collar Jobs	Workers	Professions	Senior Government Officials
I	17 (51.5)	—	—	5 (15.1)	3 (9.1)	4 (12.1)	5 (15.1)
II	—	10 (30.3)	—	3 (9.1)	1 (3.0)	2 (6.1)	4 (12.1)
III	—	—	6 (18.2)	—	—	6 (18.2)	—

Note : Figures in the parenthesis shows percentages.

The third configuration was comprised of landlords and professionals and accounted for 18.2 industrial families. In this configuration, only one type of subsidiary occupations existed.

The second aspect of occupational structure examined is the nature of non-manufacturing occupations into which the members of the industrial families had ventured after their entry into the manufacturing industry. Two major findings are :

1. Over 80 per cent members of industrial families have always been in business, trade and industry since they entered industry (Table 7.6).

TABLE 7.6 : **Participation of Family Members in Different Occupations**

Type of Occupation	Frequency and Percentage of Participating Members					
	Karta III		*Karta* II		*Karta* I	
	Frequency	Percentage	Frequency	Percentage	Frequency	Percentage
1. Business, trade and manufacturing	1963	84.2	1549	84.8	976	81.2
2. Law	108	4.6	80	4.4	46	3.8
3. Medicine	—	—	41	2.2	62	5.1
4. Engineering	—	—	29	1.6	46	3.8
5. Consultancy	261	11.2	99	5.4	53	4.4
6. Bureaurcacy	—	—	29	1.6	21	1.7
Total membership	2332	100.0	1827	100.0	1204	100.0

2. The remaining 20 per cent or so members had taken up occupations that were complementary to the manufacturing process. However, at different points of time, varying emphasis appeared to have been given to these occupations by the industrial families. Choice of law as a profession has remained almost static whereas consultancy has gained wider acceptance over the years. It is notable that the major users of consultancy services are none other than the industrial families themselves.

DEVELOPMENT OF ENTREPRENEURIAL CAPABILITIES IN FAMILY MEMBERS

Discussions with the respondents indicated that the industrial families also handle the function of training their adults along with the much discussed process of socialization of their children. Each male member of the family is watched vis-a-vis his special talents, aptitude and aspirations by the members of the core councils and the *kartas* and, as they grow up, are groomed in stages for specific roles in the economic enterprises of the families. These entrepreneurial stages are :

(a) apprenticeship : under a wide variety of individuals some of whom may actually be dealing with non-strategic functions ;

(b) co-authorship : under the guidance of a senior member of the core council who makes strategic decisions ;

(c) partial autonomy : handling of a few strategic functions within the guidelines provided ; and

(d) fuller autonomy : handling of a larger range of functions in the assigned area.

This type of entrepreneurial training has the following advantages :

1. Since strategic decisions are being taken at the seat of family management, learning is not confined to the work environment alone. It is extended during all the waking hours, i.e., at breakfast, lunch, dinner tables, at social gatherings and at other ceremonial occasions. Thus, learning takes place in natural surroundings and assimilation is quick.

2. In this system the learner and his role-model generally operate in close concert and as such the element of error in exercising options is quickly sorted out before it can do any appreciable damage.

3. Learning itself is devoid of theoretical paddings. On the contrary, it is down to earth and very much interwoven with family traditions.

4. Before the assumption of fuller autonomy the learner is being controlled by someone who is his family member and whose writ always carries weight over the heads of the salaried

executives howsoever highly placed they may be.

As a result of the operation of this adult socialization process, the industrial families have built up a reservoir of entrepreneurial talent in their midst. This vests in them larger capabilities which, coupled with their superior resource base, are utilized in keeping others at a disadvantage in entering a new line, surviving in the midst of stiff competition in the existing ones, and maintaining a prominent position. No regulatory mechanism can possibly stop these advantages.

SATISFACTION OF NEEDS OF THE FAMILY MEMBERS

The industrial families fulfil the following needs of their family members through their development and deployment in economic enterprises :

1. Identity

The identity needs of the family members are fulfilled in two ways : (i) by being members of the corps of the family engaged in managing the family enterprise and (ii) by affiliation with one or more number of economic units. These affiliations take care of the identity question vis-a-vis society at large and specific segments of the business environment (both internal and external). The members of the family take pride in the accomplishments of the units with which they are affiliated. Legends grow inside and outside the families about the manner in which the concerned members sort out their turbulence. It also develops in them a sense of loyalty to their units among the larger comity of units controlled by their families.

2. Status

As indicated earlier, different family members are distributed in different strata of management structures. This hierarchy confers status in proportion to the position occupied in the management structure. The top position goes to the *karta* who heads the core council. The next important positions are those of the core council members. The third layer consists of members of the substructures, followed by all others figuring on the unit level managements without any placement on the superstructures. For the last category of these family members, another determinant of status is the rank and importance of the unit(s) on which they are placed and their personal positions (i.e., director, fulltime/ executive director, deputy/joint managing director, chairman/ managing director, chairman/president). Almost all industrial families have a ranking of their economic units and, as such, placements on important units confer a higher status.

Status gains are dovetailed with experience and performance, and within the families competition is always on among the younger set of members to gain ever higher status. In this race no quarter is asked for and none is given. The family authority structures, however, make it a point to reward a good performance in order to encourage those who have done well in their assigned territories.

3. Role and participation in decision-making

Since most strategic decision-making in the industrial families is made at the superstructure level, and most of the individual family members are identfiied with the units, the satisfaction of their personal ego is a key issue. The industrial families have solved it through the use of following mechanisms :

(a) members placed on the concerned units are always consulted and involved in strategic decision-making which shapes the course of those units ;

(b) once the strategic decisions are taken, their implementation is via the concerned members ;

(c) the *karta*, core councils and the substructures generally do not interfere in the day-to-day administrative and operative decisions ; and

(d) non-members of the superstructure are given freedom to work out their ideas and bring it to the top. If they can sell them, the superstructure accepts them, giving full marks to the initiators of the proposal. There are numerous examples from the industrial families where some interesting projects have germinated via this route.

4. Prestige

Family members get prestige via their performance of assigned chores. Thus, prestige generally follows the successful execution of a commissioned job, erection of a plant, introduction of a new product, expansion, etc. The greater the task and the stakes involved, the higher is the prestige. All the industrial families have a mixed bag of the able, the average and the not so effective members. As such, they tend to have some "trouble-shooters" who can take on additional responsibilities of management of turbulence. Such individuals enjoy a uniformly higher prestige in all the industrial families. Those who do not do too well in some of their assignments are switched over to comparatively less demanding tasks or units. To develop their confidence, they are gradually pushed up again.

5. Reward

The reward system operative in the industrial families takes the following two forms : it enhances the status of the rewarded persons with a promotion in the hierarchy, or a more prestigious assignment ; and it fulfils some personal cherished desire (i.e. a sports car, a new house, a project of their choice) of the concerned member.[3]

The reward system is, however, not a rationalized mechanism. Accomplishment of assigned tasks needs the support of resources (men, money, material and liaison). In quite a few cases, it was revealed, that some family members had received greater indirect support to achieve their results and consequent rewards than others. Generally, these family members were the ones in whom the *karta* and other core council members took special interest. Thus, the entire process of satisfaction of the ego needs of the family members is a highly personalized system, and one in which specificities rather than generalities decide the choices. In its operation it works very satisfactorily for some and not so well for certain others. However, for the family as a whole corporate unit, the system works fairly well.

PRESERVATION OF THE CORPORATE ENTITY OF THE FAMILIES

Earlier, it was noted that the industrial families have displayed a very high rate of turbulence of various kinds. However, it was also indicated that barring one sample family, all others have succeeded in preserving their corporate entities. Whereas every type of turbulence does disturb the prevailing equilibrium in the families, those that directly threaten their unity and existence relate to succession, and to the use of economic units for personal benefit and at the cost of the family.

Succession to the position of *karta* always is a traumatic experience in every family. What is at stake here is not the share in the family property. It is the power and authority to preside over the destinies of family empires. The industrial families have handled the succession issue in the following manner :

1. From amongst the core council members, one individual is generally picked up to succeed the *karta* at a fairly early stage itself. The potential *karta*, therefore, is groomed under the personal supervision of the reigning *karta*. Thus, the family and business environment is notified about the intentions, as it were, to accept the designated person. While the succession itself actually has not taken place, the reaction of the other members of the family and segments of business environment to the impending event becomes known to the concerned families. This offers opportunities to all concerned to work out the gradual shifting of responsibilities, and sort out transitional problems.

2. In the industrial families, the choice of successors is governed by familial traditions. These traditions are not drastically different from the general Indian practice where the eldest son succeeds the father. However, the succession issue becomes complicated as the families become large and more complex. As Table 7.7 clearly demonstrates, the prospect of the eldest son of the previous *kartas* succeeding has gradually been declining. Where there has been a departure from the eldest son succeeding, the largest percentage of cases, the successor expectedly is the senior-most brother of the previous *karta*. However, there have been instances of even junior brothers and junior sons succeeding under special circumstances where the next senior person in either generation was either not acceptable to the majority in the family or was mentally or physically not equipped to do so.

One reason why the industrial families continue to be single corporate entities is that it gives them distinct economic advantages to remain so. Remaining together not only minimizes their family expenses, it gives them tax concessions and confers extra resourcefulness in exploiting business opportunities. However, this is the case only so long as all the family members adhere to the family traditions and values and do not sacrifice the interests of the family at the altar of personal gains. Once

TABLE 7.7 : **Succession in the Industrial Families**

Type of Succession	Frequency and Percentage of Families					
	Karta III		*Karta* II		*Karta* I	
	Frequency	Percentage	Frequency	Percentage	Frequency	Percentage
1. Eldest son of the previous *karta*	21	63.6	22	66.7	25	75.7
2. Next seniormost brother of previous *karta*	5	15.1	4	12.1	3	9.1
3. Previous *karta's* brother but not seniormost	5	15.1	3	9.1	1	3.0
4. Previous *karta's* son but not eldest	2	6.1	3	9.1	4	12.1
5. Adopted son	—	—	1	3.0	—	—

individuals start building personal fortunes in the process of which the family interests start suffering, a very serious threat to the familial unity springs up.

It would be incorrect to state that no such crises have occurred among the industrial families. On the contrary, almost every family has experienced such crises. In fact, except in cases where the family itself has decided to make some of its units sick in order to gain in the short run, or where the market and environmental conditions makes a unit sick, in most other cases sickness has been caused by such crises where one or two individuals from the family have milked the units dry to build their personal fortunes.

The response of various families to such crises has been varied. Some families have promptly clipped the wings of their erring members by taking 'legal' measures that is, by reducing them to inconsequential decision-making positions. Certain other families, where legal measures could not be taken because of some reason or the other, have helplessly watched the ruin of part of their economic empires. However, they have taken care to keep such "culprits" out from any new ventures which they may have started, and within the family, although no formal split has taken place, such individuals are treated more or less as *persona non grata.*

DEVELOPMENT AND DEPLOYMENT OF ECONOMIC RESOURCES

Development and deployment of economic resources involves management of own wealth and managed wealth, separately and in combination. In the following analysis, estimated own wealth, its deployment in various areas, sequence and nature of deployment of own wealth are first presented ; estimated managed and total wealth of the industrial families are considered next ; and proportions of own wealth in total wealth and managed wealth are dealt with thereafter.

The estimates of own, managed and total wealth, and deployment of own wealth in various areas were obtained from the third and, in some cases, second sets of *kartas* (now retired). It is essential to point out the basic difference between the data on economic resources of this study and that of sources such as the Department of Company Affairs, the Reserve Bank of India, and the *Economic Times*. In the present study, own and managed wealth and total wealth in the case of each family are considered as separate conceptual categories, whereas the later mentioned sources are generally concerned with the last two items and do not go into the details of own wealth. In so far as the reliability of the figures given are concerned, figures from none of the sources are beyond doubt. In the present case, the data were obtained from primary sources. Since their identity was guaranteed not to be disclosed, it is reasonable to assume that the data are nearer the truth.

Estimated own wealth

The range of estimated own wealth varied between Re. 1 and Rs. 30 crore at the time of the first set of *kartas*, between Re. 1 and Rs. 50 crore at the time of second set of *kartas* and between Rs. 11 and Rs. 70 crore at the time of present set of *kartas* (Table 7.8) and Diagram 7.1). The spread of the range of size of own wealth was thus quite narrow at the time of the first and second sets of *kartas*, it has shown much width at the time of the present set of *kartas*. Also evident is the fact of somewhat more normalized distribution of the industrial families among various size intervals at the time of the present set of *kartas* as against skewed concentration among the lowest two class intervals at the time of the first and second sets of *kartas*. The average size of own

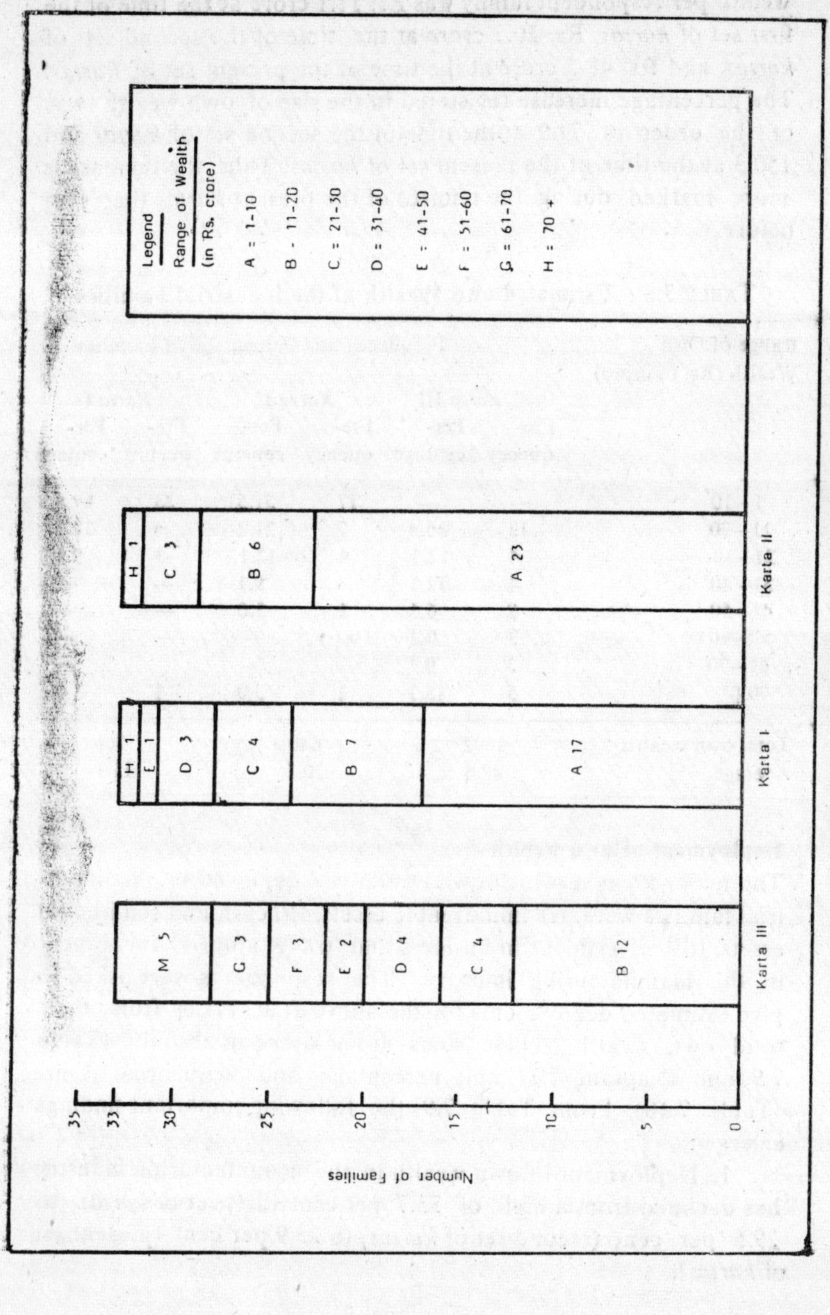

FIG. 7.1 : *Range of own wealth of Industrial Families*

wealth per respondent family was Rs. 11.1 crore at the time of the first set of *kartas*, Rs. 20.7 crore at the time of the second set of *kartas*, and Rs. 48.5 crore at the time of the present set of *kartas*. The percentage increase registered in the size of own wealth was of the order of 74.9 at the time of the second set of *kartas* and 150.3 at the time of the present set of *kartas*. Thus, the increase is more marked during the tenures of the present *kartas* than ever before.

TABLE 7.8 : **Estimated own Wealth of the Industrial Families**

Range of Own Wealth (Rs. in crore)	Frequency and Percentage of Families					
	Karta III		*Karta* II		*Karta* I	
	Frequency	Percentage	Frequency	Percentage	Frequency	Percentage
1—10	—	—	17	51.5	23	69.7
11—20	12	36.4	7	21.2	6	18.2
21—30	4	12.1	4	12.1	3	9.1
31—40	4	12.1	3	9.1	—	—
41—50	2	6.1	1	3.0	—	—
51—60	3	9.1	—	—	—	—
61—70	3	9.1	—	—	—	—
70+	5	15.1	1	3.0	1	3.0
Total own wealth	1602		640		366	
Average	48.5		20.7		11.1	

Deployment of own wealth

The major areas in which own wealth was deployed by the industrial families were, (i) immoveable assets, (ii) cash and semi-liquid assets, (iii) investment in business and trade and (iv) investment in the manufacturing industry. The respondents were asked to give estimated deployments for the above four areas from their total own wealth. These sums are analysed in absolute (Table 7.9 and Diagram 7.2) and percentage and sequential terms (Table 7.10). From Table 7.9 the following important findings emerge :

1. Deployment of own wealth in the manufacturing industry has declined from a high of 33.7 per cent (first set of *kartas*) to 29.5 per cent (second set of *kartas*, to 25.9 per cent (present set of *kartas*).

TABLE 7.9 : **Own wealth and its deployment in various areas**

Deployment of own Wealth in Various Area (Rs. in crores)	Frequency and Percentage of Deployment					
	Karta III		*Karta* II		*Karta* I	
	Fre-quency	Per-centage	Fre-quency	Per-centage	Fre-quency	Per-centage
1. Deployment in immovable assets	307.4	19.2	110.9	17.3	64.8	17.7
2. Deployment in cash, semiliquid assets	546.3	34.1	155.6	24.3	73.3	20.0
3. Investment in business and trade	334.0	20.8	184.5	28.8	104.5	28.6
4. Investment in manufacturing industry	414.3	25.9	189.0	29.5	123.4	33.7
Total own wealth	1602	100.0	640	100.0	366	100.0

2. Deployment of own wealth in business and trade had remained around 28 per cent during the tenures of the first and second set of *kartas* but has declined to 20.8 per cent at the time of present set of *kartas*.

3. Keeping of cash and semi-liquid assets for ready use in the pursuit of various strategies chalked out has increased very appreciably from 20 per cent (first set of *kartas*), to 24.3 per cent (second set of *kartas*), to 34.1 per cent (present set of *kartas*.)

4. Deployment of own wealth in immovable assets has also appreciated from 17.7 per cent to 19.2 per cent, but not too sharply. The sums, present in Table 7.9, were then converted into percentages. These percentages were further assigned the following qualitative labels as follows : Very High (VH) : over 30 per cent ; High (H) : 21 per cent to 30 per cent ; Moderate (M) : 11 per cent to 20 per cent ; Low (L) : 6 per cent to 10 per cent and Very Low (VL) : below 6 per cent. Arranged on the time continuum in a sequence in the case of all the respondent families, the data yielded 10 different entrepreneurial mixes of own wealth :

1. Moderate assets, investment in business trade and manufacturing industry, and very high cash at the time of the first set of *kartas*, moving to moderate assets and cash and high investment in business, trade manufacturing at the time of the second

set of *kartas* ; and moderate assets and cash, and investment in business and trade, and high investment in manufacturing industry at the time of the present set of *kartas*—18.2 per cent of families are of this type.

2. High assets and cash and moderate investments in business, trade and manufacturing industry at the time of the first set of *kartas*; moving to moderate assets, investment in business, trade and manufacturing industry, and very high cash at the time of the second and third sets of *kartas*—15.1 per cent of families belong to this type.

3. High assets and cash and moderate investments in business, trade and manufacturing industry at the time of the first set of *kartas* ; moving to moderate assets, investment in business, manufacturing industry and high cash, and investment in business and trade at the time of the second set of *kartas* ; and moderate assets and cash, and high investment in business, trade and manufacturing industry at the time of the third set of *kartas*—9.1 per cent families constitute this pattern.

4. Moderate assets and high cash, and investment in business, trade and manufacturing industry at the time of the first set of *kartas* ; moving to low assets and high deployment in the remaining three areas at the time of the second and third sets of *kartas*—6.1 per cent families belong to this pattern.

5. Low assets, high cash, moderate investment in business and trade and very high investment in the manufacturing industry at the time of the first set of *kartas* ; moving to low assets, moderate cash, high investment in business, trade and very high investment in manufacturing industry at the time of the second set of *kartas* ; and low assets, moderate cash, investment in business, trade and very high investment in manufacturing industry at the time of the third set of *kartas*—12.1 per cent families comprise this pattern.

6. Moderate assets and cash, and high investment in business, trade and manufacturing industry during the reigns of the first set of *kartas* ; moving to low assets, moderate cash, investment in business and trade, and very high investment in manufacturing industry at the time of the second set of *kartas* ; and low assets, moderate cash, and high investment in business and trade, and very high investment in manufacturing industry at the time of the third set of *kartas*—9.1 per cent families are of this type.

7. Very low assets, very high cash, high investment in business and trade, and moderate investment in the manufacturing industry at the time of the first set of *kartas* ; moving to low assets and high deployment in the remaining three areas at the time of the second and third sets of *kartas* ; three per cent families belong to this pattern.

8. Very low assets, high cash, very high investment in business and trade, and moderate investment in manufacturing industry at the time of the first set of *kartas* ; moving to moderate assets and high deployment in the remaining three areas at the time of the second set of *kartas* ; and moderate assets, very high cash and high investment in business, trade and manufacturing industry at the time of the third set of *kartas*—6.1 per cent families are of this type.

9. High assets, very high cash, moderate investment in business and trade, and high investment in manufacturing industry at the time of the first set of *kartas* ; moving to moderate assets, very high cash and high investment in business, trade and manufacturing industry at the time of the second set of *kartas* ; and moderate assets, high cash, very high investment in business, trade and manufacturing industry at the time of third set of *kartas*—12.1 per cent of families belong to this pattern.

10. High assets, very high cash, moderate investment in business and trade, and high investment in manufacturing industry at the time of first set of *kartas* ; moving to high assets, moderate cash, very high investment in business, trade, and high investment in manufacturing industry at the time of the second set of *kartas* ; and high assets, moderate cash and very high investment in business, trade and manufacturing industry at the time of the third set of *kartas*—9.1 per cent families belong to this pattern.

The foregoing patterns reflect the entrepreneurial shifts in emphasis on different components of deployment mix at the time of various set of *kartas*. To the extent that deployment of own wealth in one area came down, say, from very high to moderate level, it meant a change in the complexion of the resource mix used.

Estimated managed wealth

Estimated managed wealth means total wealth minus own wealth of the industrial families. This wealth is a mix of contributions made by shareholders of economic units, investments made by term-lending institutions, commercial banks and, in a few cases, even the state or central governments directly. The range of estimated managed wealth among the industrial families varied between Rs. 1 crore and Rs. 150 crore at the time of the first set of *kartas*, between Rs. 21 crore and over Rs. 250 crore at the time of the second and third sets of *kartas* (Table 7.11 and Diagram 7.3). The distribution of the industrial families among the lowest three class intervals of the range of estimated managed wealth was quite concentrated during the tenures of the first and second sets of *kartas*; it is evenly distributed at the time of the present set of *kartas*. Per family averages of estimated managed wealth worked out to Rs. 15.9 crore (first set of *kartas*), Rs. 35.1 crore (second set of *kartas*) and Rs. 148.2 crore (present set of *kartas*). The increase registered in the estimated managed wealth of the industrial families was of the order of 120 per cent during the tenures

TABLE 7.11 : **Estimated managed wealth* of industrial families**

Range of Managed Wealth (Rs. in crores)	Frequency and Percentage of Families					
	Karta III		*Karta* II		*Karta* I	
	Frequency	Percentage	Frequency	Percentage	Frequency	Percentage
1— 10	—	—	5	15.1	17	51.5
11— 20	—	—	11	33.3	8	24.2
21— 30	3	9.1	6	18.2	6	18.2
31— 40	3	9.1	1	3.0	1	3.0
41— 50	3	9.1	4	12.1	—	—
51— 60	3	9.1	4	12.1	—	—
61— 80	2	6.1	1	3.0	—	—
81—100	2	6.1	—	—	—	—
101—150	5	15.1	—	—	1	3.0
151—200	6	18.2	—	—	—	—
201—250	3	9.1	—	—	—	—
250+	3	9.1	1	3.0	—	—
Total managed wealth	4890		1159		525	
Average	148.2		35.1		15.9	

*Estimated Managed Wealth means Total Wealth minus Own Wealth.

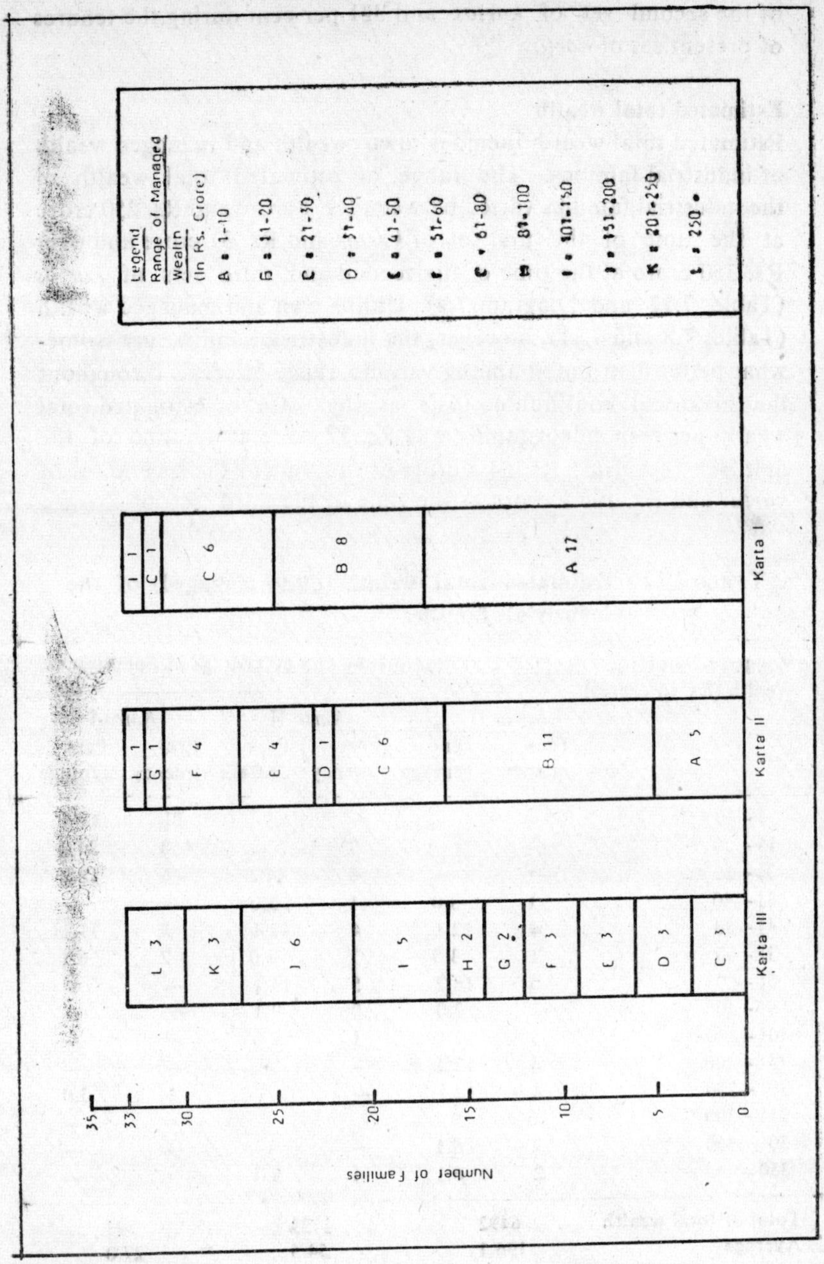

FIG. 7.3 ; *Range of managed wealth of Industrial Families*

of the second set of *kartas* and 321 per cent during the tenures of present set of *kartas*.

Estimated total wealth

Estimated total wealth includes own wealth and managed wealth of industrial families. The range of estimated total wealth of the industrial families varied between Rs. 1 crore and Rs. 250 crore at the time of the first set of *kartas* and Rs. 31 crore and over Rs. 350 crore at the time of the second and third sets of *kartas* (Table 7.12 and Diagram 7.4). Unlike own and managed wealth (Tables 7.8 and 7.11), however, the industrial families are somewhat better distributed among various range intervals throughout the historical continuum. The average size of estimated total wealth per respondent family was Rs. 27 crore at the time of the first set of *kartas*, Rs. 54.5 crore at the time of the second set of *kartas* and Rs. 196.1 crore at the time of the third set of *kartas*.

TABLE 7.12 : **Estimated total wealth (own+managed) of the industrial families**

Range of Total Wealth (Rs. in crores)	Frequency and Percentage of Families					
	Karta III		*Karta* II		*Karta* I	
	Fre-quency	Per-centage	Fre-quency	Per-centage	Fre-quency	Per-centage
1—10	—	—	—	—	11	33.3
11—20	—	—	9	27.3	8	24.2
21—30	—	—	8	24.2	6	18.2
31—40	1	3.0	1	3.0	—	—
41—50	4	12.1	4	12.1	5	15.1
51—60	1	3.0	1	3.0	2	6.1
61—80	6	18.2	5	15.1	—	—
81—100	1	3.0	3	9.1	—	—
101—150	5	15.1	1	3.0	—	—
151—200	4	12.1	—	—	—	—
201—250	4	12.1	—	—	1	3.0
251—300	3	9.1	—	—	—	—
301—350	2	6.1	—	—	—	—
350+	2	6.1	1	3.0	—	—
Total of total wealth	**6492**		**1799**		**891**	
Average	**196.1**		**54.5**		**27.0**	

The order of increase registered in the estimated total wealth of industrial families was 101.9 per cent at the time of the second set of *kartas*, and 260.9 per cent at the time of the third set of *kartas*. Juxtaposing the increase registered in own, managed and total wealth of the industrial families, it emerges that the highest percentage increases are registered by total wealth followed by managed wealth and own wealth, in that order.

In discussing the three types of wealth of the industrial families, two important dimensions are : the proportion of three types of wealth and the proportion of own wealth in total and managed wealth. Tables 7.13, 7.14 and 7.15 present these data. The proportion of own wealth had declined from a high of 41.2 per cent at the time of the first set of *kartas*, to 35.6 per cent at the time of the second set of *kartas*, to 24.6 per cent at the time of the present set of *kartas* (Table 7.13). The percentage of managed wealth had correspondingly increased from a low of 58.8 (first set of *kartas*), to 64.4 (second set of *kartas*), to 75.4 (third set of *kartas*). However, the proportion of managed wealth has always been substantially higher among the industrial families. The position at present is that the ratio between own and managed wealth stands at 3 : 1. This tendency is, however, not surprising since greater utilization of others' wealth for running their economic enterprises is a trait found among industrialists all over the world.

The range of proportions of own wealth in total wealth varied between 15 per cent and over 45 per cent at the time of the first and second sets of *kartas*, and between 15 per cent and 45

TABLE 7.13 : **Proportion of own and managed wealth in total wealth of the industrial families**

Composition of Total Wealth	Proportion from Total Wealth					
	Karta III		*Karta* II		*Karta* I	
	Frequency	Percentage	Frequency	Percentage	Frequency	Percentage
1. Own wealth	1602	24.6	640	35.6	366	41.2
2. Managed wealth	4890	75.4	1159	64.4	525	58.8
Total wealth	6492	100.0	1799	100.0	891	100.0

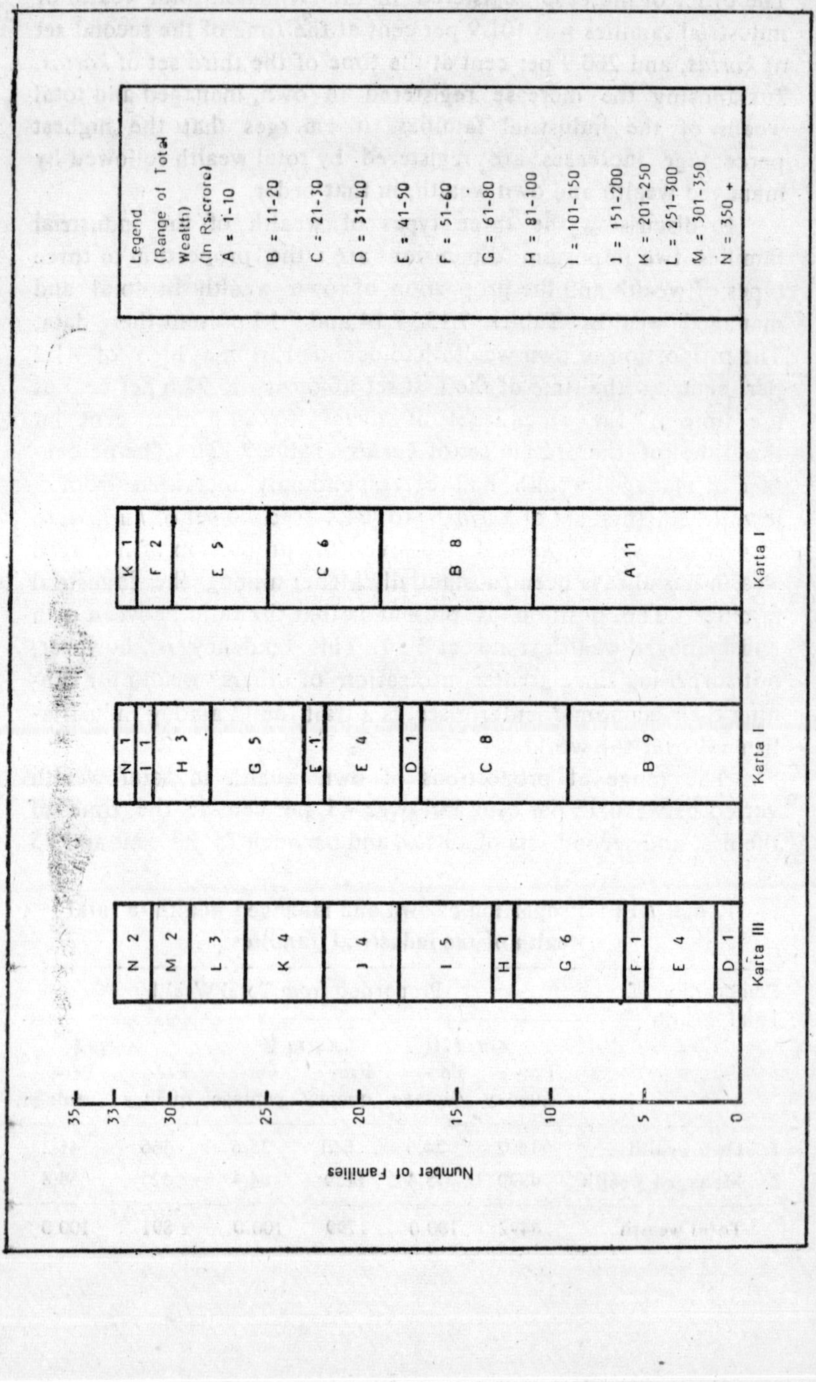
Number of Families
0
5
10
15
20
25
30
33
35
Karta III
D 1
E 4
F 1
G 6
H 1
I 5
J 4
K 4
L 3
M 2
N 2
Karta I
B 9
C 8
D 1
E 4
E 1
G 5
H 3
I 1
N 1
Karta I
A 11
B 8
C 6
E 5
F 2
K 1
Legend
(Range of Total
Wealth)
(In Rs.crore)
A = 1-10
B = 11-20
C = 21-30
D = 31-40
E = 41-50
F = 51-60
G = 61-80
H = 81-100
I = 101-150
J = 151-200
K = 201-250
L = 251-300
M = 301-350
N = 350 +

TABLE 7.14 : **Proportion of estimated own wealth in total wealth in the industrial families**

Range of Proportion of own wealth (in percentages)	Frequency and Percentage of Families					
	Karta III		*Karta* II		*Karta* I	
	Fre-quency	Per-centage	Fre-quency	Per-centage	Fer-quency	Per centage
15-20	6	18.2	1	3.0	1	3.0
21-25	9	27.3	4	12.1	3	9.1
26-30	12	36.4	6	18.2	5	15.1
31-35	5	15.1	11	33.3	4	12.1
36-40	—	—	3	9.1	8	24.2
41-45	1	3.0	6	18.2	6	18.2
45+	—	—	2	6.1	6	18.2
Average		24.6		35.6		41.2

per cent at the time of the third set of *kartas* (Table 7.14). However, it is evident that the proportion has steadily been declining over the years. In terms of linearity of decline, the proportion of own wealth in managed wealth too follows the same path (Table 7.15). Nevertheless, the actual percentage ranges are much wider (15% and over 80% as against 15% and over 45%), and the decline is more marked at the time of the present set of *kartas*.

To sum up, as a class the industrial families have displayed a strong tendency to use their human and economic resources in a manner that enabled them minimize their costs and maximize their gains. In terms of strategic decision making, this is reflected in their mixed build-up of educational and professional background of the family members, consolidation of class interest through the help of marriages with their allies in society, and a general concentration of their members in business, trade and manufacturing occupations. To keep their hold on an expanding economic empire, they build a reservoir of entrepreneurial talent within their families via the adult socialization process, satisfying the identity needs of their members, and presering their corporate entities. Their deployment of own wealth in different mixes of investment components also is geared towards producing maximum possible benefits.

TABLE 8.1 : **Total number of economic units with the industrial families**

Range of Number of Units	Frequency and Percentage of Families					
	Karta III		*Karta* II		*Karta* I	
	Frequency	Percentage	Frequency	Percentage	Frequency	Percentage
1- 10	8	24.2	17	51.5	24	72.7
11- 20	12	36.4	5	15.1	3	9.1
21- 30	2	6.1	4	12.1	—	—
31- 40	4	12.1	1	3.0	3	9.1
41- 50	1	3.0	1	3.0	—	—
51- 60	—	—	1	3.0	2	6.1
61- 70	2	6.1	1	3.0	—	—
71- 80	1	3.0	1	3.0	—	—
81- 90	1	3.0	1	3.0	—	—
91-100	—	—	—	—	—	—
101-150	1	3.0	—	—	—	—
150+	1	3.0	1	3.0	1	3.0
Total number of units	1262		900		533	
Average	38.2		27.3		16.1	

average per family of units came to 27.3. The range of the number of units did not change much during the tenures of the third set of *kartas*, but the distribution of families in various range intervals did so. As against the concentration of 72.7 per cent and 51.5 per cent families in the interval 1 unit and 10 units at the time of the first and second sets of *kartas*, respectively, the percentage of industrial families in the same range interval had dropped to 24.2. Similarly, the next range interval, namely, 11 units and 20 units, had as much as 36.4 per cent families, with the remainder distributed among the other 10 class intervals. The average per family went up to 38.2. The increase registered in the number of economic units with the industrial families was 68.9 per cent at the time of the second set of *kartas* and 40.2 per cent at the time of the third set of *kartas*.

The largest number of new units were established or acquired at the time of the first set of *kartas*. The same tempo was not maintained thereafter (Table 8.2). Nor was it expected to be

TABLE 8.2 : **Number of new economic units with the industrial families**

Range of Number of New Units	Frequency and Percentage of Families					
	Karta III		*Karta* II		*Karta* I	
	Fre-quency	Per-centage	Fre-quency	Per-centage	Fre-quency	Per-centage
0	—	—	1	3.0	—	—
1- 5	16	48.5	15	45.4	19	57.6
6-10	13	39.4	6	18.2	5	15.1
11-15	1	3.0	3	9.1	2	6.1
16-20	—	—	2	6.1	1	3.0
21-25	—	—	2	6.1	—	—
26-30	2	6.1	2	6.1	—	—
30+	1	3.0	2	6.1	6	18.2
Total number of new units	362		367		533	
Average	11.0		11.1		16.1	

maintained, since the subsequent task for the industrial families was to run and manage the old units. Predictably, the largest percentages of families were concentrated in unit class intervals of 1 and 5 and 6 and 10. The average number of new units established per family had remained unchanged at 11.0 since the time of the second set of *kartas*.

Increasing capacities via expansion and diversifications indicate the directions of growth consciously preferred by the industrial families. It is evident that not many major expansions were attempted by them (Table 8.3). Over 80 per cent families reported major expansions or diversifications numbering up to six at the time of the first set of *kartas*. This position deteriorated still further during the subsequent period. The average number of expansions or diversifications stood at 6.3, 6.0, and 2.2 at the time of first, second and third set of *kartas*, respectively. The respondents identified the following factors that were responsible for this occurrence.

1. A general practice among Indian large-scale manufacturers is to install capacity much in excess of that licensed. Therefore, the task before the new units quite often is that of capacity utilization.

TABLE 8.3 : **Number of major expansions/diversifications effected**

Range of Expansions/ Diversifications	Frequency and Percentage of Families					
	Karta III		*Karta* II		*Karta* I	
	Frequency	Percentage	Frequency	Percentage	Frequency	Percentage
0	3	9.1	—	—	—	—
1	9	27.3	2	6.1	—	—
2	9	27.3	2	6.1	1	3.0
3	6	18.2	8	24.2	5	15.1
4	3	9.1	7	21.2	4	12.1
5	2	6.1	2	6.1	6	18.2
6	1	3.0	4	12.1	8	24.2
7	—	—	—	—	2	6.1
8	—	—	—	—	2	6.1
9	—	—	1	3.0	1	3.0
10	—	—	2	6.1	1	3.0
10-15	—	—	3	9.1	1	3.0
16-20	—	—	2	6.1	2	6.1
Total expansions/ diversifications		**30**		**198**		**209**
Average		**2.2**		**6.0**		**6.3**

2. A major expansion also generally involves technological choices, and since technology is to be purchased, it necessitates additional liaison work with the state apparatus. The economic costs of these alternate technologies are sometimes prohibitive.

3. Major expansions are costly operations. In quite a few industries it is cheaper to install a new unit rather than expand the capacity of an existing one.

The general operation of an entrepreneurship process in the Indian large-scale manufacturing sector has been : entry→growth →consolidation→dominance→diversification—. This process has operated in the case of every single product in which an entrepreneur has begun his career. Thus, diversification is inherent in the process of the growth of enterprises.

Although the number of economic units with some of the

industrial families was fairly large, the coverage of products or product groups was not so (Table 8.4). The average number of products or product groups covered at different stages were : 3.2 (first set of *kartas*), 6.5 (second set of *kartas*), and 8.7 (third set of *kartas*). The product coverage of two-thirds of the industrial families was confined only to two products or product groups at the time of the first set of *kartas*. Gradually, however, the range of coverage has increased, and the number of products covered has touched as high a figure as 30 and over during the tenures of the second and third sets of *kartas*. The increase was of the order of 104.8 per cent at the time of the second set of *kartas* and 33 per cent at the time of the third set of *kartas*. Two other notable aspects of this coverage are : first, with a few exceptions, the industrial families had not established many heavy and core sector industries[1] and largely, had concentrated on mass consumption consumer items ; second, research and

TABLE 8.4 : **Number of products covered**

Number of Products Covered	Frequency and Percentage of Families					
	Karta III		*Karta* II		*Karta* I	
	Frequency	Percentage	Frequency	Percentage	Frequency	Percentage
1	—	—	—	—	14	42.4
2	1	3.0	4	12.1	7	21.2
3	3	9.1	11	33.3	4	12.1
4	5	15.1	3	9.1	—	—
5	6	18.2	4	12.1	1	3.0
6	2	6.1	3	9.1	2	6.1
7	5	15.1	—	—	1	3.0
8	1	3.0	1	3.0	1	3.0
9	2	6.1	2	6.1	1	3.0
10	1	3.0	—	—	1	3.0
11-15	2	6.1	2	6.1	1	3.0
16-20	2	6.1	1	3.0	—	—
21-25	2	6.1	1	3.0	—	—
26-30	—	—	—	—	—	—
30+	1	3.0	1	3.0	—	—
Total products covered	286		215		105	
Average	8.7		6.5		3.2	

development effort was almost totally absent. As a result, the technological base of most of the units operated by them has become obsolete.

WORKERS

Quite a few theories of wage determination—marginal productivity, collective bargaining, wage differentials, etc.—have been enunciated. These theories quite often do not point very clearly to the structurally unequal nature of the relationship existing between employers and employees. What needs to be recognized is the inherently greater advantage vested in employers in their dealing with workers by their larger capacity to control labour and greater freedom to choose their options. What further increases the dependency of the workers are, (i) very low rates of employment, (ii) greater flexibility and mobility of capital and (iii) the need of the workers to work in order to survive.[2]

The structure of the exchange relationship is such that the dominant or controlling authority—the employers—and the subordinates—the workers—occupy the two ends of the continuum. They work towards maximizing their control and reducing domination respectively. In reality, however, the relationship is made very complex because of the fluctuating role of the state as a mediator in determining various aspects of this relationship.

In analysing the structure of the exchange relationship between the industrial families and the workers employed by them, the present study uses a slightly modified version of framework provided by Crouch (1977, pp. 14-25). This framework has three main aspects, (i) structure of inequality in wage determination, (ii) style of exercising dominance and (iii) response to the exercise of countervailing power by the subordinates. The first of these is considered in isolation, the last two are analysed in combination. The data on these items provides the attitudinal justification for the behaviour of industrial families inside their economic units and in negotiating with elements outside their environment.

The structure of inequality in wage determination reflects the normative orientation of the industrial families. It finds its expression in the steps that they take in wage negotiation, actual determination of wage, wage payment and continuation of low wage structure. The following are its major aspects, (i) justification of inequality (wage inequality justified on moral or religious grounds or as an objective reward system), (ii) causes of inequality (caused by administrative decisions or socio-economic interactions), (iii) approaches to wage determination (systematic or *ad hoc*) and (iv) growth of enterprises (at the expense of others or with a fair degree of distribution of the gains).

An overwhelming majority of the industrial families—quite often ranging between two-thirds to three-fourths—justifies wage inequalities on moral or religious grounds (Table 8.5). Over the years, there has been some change in their outlook but about two-thirds among them still continue to justify wage

TABLE 8.5 : **Structure of inequality in wage determination**

Aspects and Nature of wage Determination	Frequency and Percentage of Families					
	Karta III		*Karta* II		*Karta* I	
	Frequency	Percentage	Frequency	Percentage	Frequency	Percentage
1. *Justification of Inequality*						
a. Moral/religious	22	66.7	21	63.3	24	72.7
b. Objective reward	11	33.3	12	36.7	9	27.7
2. *Causes of Inequality*						
a. Administrative decisions	26	78.8	22	66.7	20	60.6
b. Socio-economic interactions	7	21.2	11	33.3	13	39.4
3. *Approaches to Wage Determination*						
a. Systematic	3	9.1	4	12.1	3	9.1
b. Ad hoc	30	90.9	29	87.9	30	90.9
4. *Growth of Enterprises*						
a. At the expense of others	27	81.8	24	72.7	22	66.7
b. Fair distribution of gains	6	18.2	9	27.3	11	33.3

inequalities on moral or religious grounds. On the second dimension concerning the causes of inequality, the belief that it is caused by administrative decisions taken by them rather than by socio-economic interactions has been strengthened. This was evident from the increase in the percentage of families holding this view (from 60.6 in the first set of *kartas* to 66.7 in the second set of *kartas* to 78.8 in the third set of *kartas*). Wage determination as an exercise in itself is viewed as an *ad hoc* issue, to be periodically faced and tackled, by about 90 per cent families, minimum wage legislation notwithstanding. The implicit meaning hidden in the foregoing three aspects is made more clear by the percentages of the industrial families on the fourth dimension. Here two-thirds to eight-tenths of the industrial families want growth of their enterprises, even if it means growing at the expense of others.

Employment relationship, as seen in the styles of exercising dominance over the work force and in response to the countervailing power of workers operated through their trade unions, has the following aspects : (i) compliance or obedience (strict obedience or indulgence), (ii) ideological concensus (including involvement of the work-force to accept their ideological position or keeping them separate), (iii) modalities of dealing with workers (individually or as collective entities) and (iv) conflict with employers (as a clash of class interests or a matter of right of workers).

Operation of the employment relationship by the industrial families indicates that the positions taken by them are in conformity with their normative orientation discussed earlier. Compliance of obedience by the work-force was strictly enforced by over 80 per cent of the industrial families to begin with. Over the years, there has been a gradual relaxation on this issue with the result that about 62 per cent enforced strict obedience at the time of the present set of *kartas* (Table 8.6).

In order to continue their dominance, it is essential to obtain ideological consensus from workers. This is attempted in a very open manner. The top echelons of industrial bureaucracy are in any case allies of the management. The mass of the work-force is not treated at par with them and they have to be 'managed'. It is not possible for the industrial families forcefully to extract ideological consensus ; it has to be induced with very clever

TABLE 8.6 : **Employment relationship as operationalized by industrial families**

Aspects and Nature of Employment Relationship	Frequency and Percentages of Families					
	Karta III		*Karta* II		*Karta* I	
	Fre-quency	Per-centage	Fre-quency	Per-centage	Fre-quency	Per-centage
1. *Compliance of Obedience*						
a. Strict obedience	21	62.6	24	72.7	27	81.8
b. Indulgence	12	36.4	9	27.3	6	18.2
2. *Ideological Consensus*						
a. Involvement	20	60.6	24	72.7	23	69.7
b. Separation	13	39.4	9	27.3	10	30.3
3. *Dealing with Workers*						
a. Singly	18	54.5	23	69.7	27	81.8
b. Collectively	15	45.5	10	30.3	6	18.2
4. *Conflict with Employees*						
a. Clash of interest	24	72.7	21	62.6	25	75.8
b. Matter of right	9	27.3	12	36.4	8	24.2

combinations of temptation and coercion. Between 60 per cent and 72 per cent of families preferred induced ideological consensus at different points of time.

The state, through its legislation on industrial disputes, has specified the role of the trade unions as the negotiating agents of the workers. Even so, given a choice, about 81, 69 and 54 per cent of the industrial families would have preferred to deal with workers individually at the time of the first, second and third sets of *kartas*, respectively. Since it was not possible to do so, they penetrated the trade union movement and worked out strategies for 'managing' the trade unions and their leadership so as to weaken their negotiating position. An overwhelming majority of the industrial families—varying from a low of 62 per cent to a high of 75 per cent—considered the conflict between them and their workers as a clash of interests and not as a matter of the legitimate rights of their workers. It is noteworthy that in purely qualitative terms the traditions set by the industrial families in operationalizing the exchange relationship with the work-force has more or less endured over time, although in quantitative terms their occurrence on most items has somewhat been reduced.

SHAREHOLDERS

Shareholders constitute an important segment of the internal environment. Control of the shareholders—indeed of the units themselves—is managed through investment techniques which ensure management control by the families without their owning the entire stock of shares.[3] Management of shareholders by the industrial families is examined in terms of, (i) returns on their investment and (ii) extent of their involvement in the decision making in the affairs of their units. Studying the foregoing aspects purely on the basis of the number of units, posed the following problems :

1. The number of economic units to be covered for such an analysis was very large ;
2. The number of economic units themselves was no measure of the size of shareholders, since quite a few families had opted for multi-product or unit type single units rather than establishing several separate companies ;
3. Statistical information to be handled on the issue was so enormous that it was beyond the capacities of a single individual with limited resources at his disposal to do so ;
4. Available studies on the issue did not analyse the data family-wise.[4] Even otherwise, they did not cover all the units of the families figuring in the present sample.

To overcome these problems, the total stock of shareholders of the units of respondent families was considered as one flock and rating was obtained from the respondents and others about their treatment of the shareholders.[5]

In deciding the extent of returns to ordinary shareholders, as against those holding preference shares or debentures, the industrial families appear to have followed a flexible outlook. At the time of first set of *kartas*, about one-fourth of the shareholders received very high returns. Another 12.1 per cent received returns between 11 per cent and 20 per cent. However, roughly two-thirds of the shareholders received returns below 10 per cent (Table 8.7). This pattern has undergone changes during the subsequent period when the percentage of shareholders has declined in the 'very high' category (from 24.2 to 18.2) and

TABLE 8.7 : **Returns given to the shareholders**

	Type of Returns Given	Percentage of Shareholders*		
		Karta III	*Karta* II	*Karta* I
1.	Very high (above 20 per cent)	18.2	18.2	24.2
2.	High (11-20 per cent)	27.3	21.2	12.1
3.	Moderate (6-10 per cent)	15.1	21.2	27.3
4.	Low (4-5 per cent)	21.2	18.2	18.2
5.	Very low (up to 3 per cent)	21.2	18.2	18.2

*Actual strength not indicated by the respondents.

increased in the high category (from 12.1 to 21.2 to 27.3). The proportion of shareholders in the 'moderate' category has also declined but has increased in case of the 'low' and 'very low' categories.

Juxtaposing the treatment meted out to a large percentage of shareholders with the number of units that the industrial families have succeeded in acquiring or establishing, the implications become very clear. It appears that the industrial families have paid a lesser amount of dividend to a larger segment of shareholders. By the same token, they should have received comparatively lesser reward, and this should have reflected in the sizes of own, managed and total wealth on the one hand, and the number of new units established and major expansions or diversifications effected on the other. What is observed instead is a generally disproportionate increase in the size of wealth and number of new units. A very witty member of an industrial family commented :

> Some units of some industrial families have a peculiar tendency of being in a very precarious financial condition quite often. As a consequence, the shareholders continue to have either a very low return or no return at all. However, through appropriate devices controlling inputs and outputs, these families continue to plough back substantial sums which avoid the tax dragnet. Small wonder, they are leapfrogging on the growth chart.

Shareholders have been treated as a 'necessary evil' by the

industrial families. They have to be 'consulted' about the decisions taken by the board of directors since the law so specifies. However, this consultation itself is reduced by the industrial families to the barest minimum in its quantitative occurrence (Table 8.8) and to a ritual in its quality. The holding of annual general meetings and the approval of the management

TABLE 8.8 : **Extent of involvement of shareholders in decision making**

Nature of Involvement	Percentage of Shareholders*		
	Karta III	*Karta* II	*Karta* I
1. Moderate	12.1	12.1	18.2
2. Low	21.2	30.3	27.3
3. Negligible	66.7	57.6	54.5

*Actual strength not indicated by the respondents.

sponsored resolutions is a mere formality. These are accomplished with consummate skill because of the controlling block of shares held by the concerned families and their associates. Special and extraordinary meetings are almost always requisitioned when the units are facing the following contingencies, (i) the courts have ordered it, (i) some other family has, via the share market, cornered a large percentage of shares and is in a position to challenge the 'controlling block' with the help of unattached shareholders and (iii) an internal fight has broken out within the family which has not been successfully combated. These occasions bring in 'proxy wars', 'stay orders', and even 'disturbances' at the company meetings.

THE SUPPORT SYSTEMS

The internal support systems include the following :

(a) Factory bureaucracy involved in personnel, material procurement, production, quality control, marketing and sales, and generally given the term of 'functional management' ;

(b) Finance and accounting covering the tasks of maintaining records for the benefits of the family on the one hand, and shareholders and tax agencies on the other hand ; and

(c) Legal aspects including shareholders' records and compliance of industrial and civil laws.

All appointees in these support systems have to be such people who are extremely loyal to the industrial families. Without loyalty, key positions are just not given. This is the tenet in practice even today when competence is slowly getting recognition. At the time of the first set of *kartas*, the industrial families had shown 36.4, 60.6 and 48.4 per cent appointees from their relations on the functional, finance and legal support systems, respectively. The corresponding percentages for the same positions from the community members were, 42.4, 36.4 and 48.4. During the subsequent period, a significant change has been effected *vis-a-vis* functional management where a majority (51.5 per cent) now consists of qualified professionals who are neither relations of the industrial families nor belong to their communities. Positions in finance, accounting and legal areas, however, continue to be given mostly to persons from the kin or community group even now. Overall dependence of the industrial families on the kin or community group among the support systems stood at 87.9 per cent at the time of the first set of *kartas*. It declined to 77.3 per cent at the time of the second set of *kartas*, and stands at 67.2 per cent at the time of the present set of *kartas*. That this dependence is still of the order of over 67 per cent indicates a conscious policy on the part of the industrial families to put their own men in key positions (Table 8.9).

Since these support systems work as transplants—they are neither part of the work force nor the management—they encounter problems of work and social identities.[6] In order to survive in economic units of the industrial families, they have to accept the superiority of the family members. The role space of these support systems has increased historically as the industrial families have built up their economic enterprises; this is inevitable since the families themselves cannot cope with the increasing work-load. However, not every one among these support systems enjoys recognition from the family managements. Since the really top positions at the super substructure levels are very few for these managers, a cut throat competitiveness among them has grown up to reach there.[7]

In India this situation is very shrewdly exploited by both the parties. The senior members of the industrial families exploit the

TABLE 8.9 : **Nature of appointees in various support systems**

Nature of Appointees	Frequency and Percentage of Appointees					
	Karta III		*Karta* II		*Karta* I	
	Fre-quency	Per-centage	Fre-quency	Per-centage	Fre-quency	Per-centage
Functional Managements						
1. Kin	460	12.1	654	24.2	582	36.4
2. Community members	1376	36.4	1146	42.4	678	42.4
3. Others	1950	51.5	900	33.3	339	21.2
Sub-total	3786	100.0	2700	100.0	1599	100.0
Finance and Accounting Systems						
1. Kin	1146	45.4	927	51.5	646	60.6
2. Community members	996	39.5	655	36.4	388	36.4
3. Others	382	15.1	218	12.1	32	3.0
Sub-total	2524	100.0	1800	100.0	1066	100.0
Legal Systems						
1. Kin	459	36.4	382	42.4	258	48.4
2. Community members	650	51.5	409	45.4	258	48.4
3. Others	153	12.1	109	12.1	17	3.2
Sub-total	1262	100.0	900	100.0	533	100.0
All Support Systems						
1. Kin	2065	27.3	1963	36.4	1486	46.5
2. Community members	3022	39.9	2210	40.9	1324	41.4
3. Others	2485	32.8	1227	22.7	388	12.1
Grand total	7572	100.0	5400	100.0	3198	100.0

managers by denying them real independence in the work sphere and by keeping them in constant fear and suspense. The managers are conscious of this reality but find themselves helpless on three counts.

1. Their remuneration and perquisites are very high in comparison to the salaries paid in other white collar jobs elsewhere ;
2. All private sector managements practise the same rules ;
3. Having worked for a long periods in the private sector, it

is difficult for them to adjust to the totally different administrative culture of the public, co-operative and joint sectors. Thus, an apparently contradictory position of a highly dissatisfied managerial cadre continuing to work persists.

In the existing situation the managers, in their efforts to make a more favourable impression on their bosses, end up in saying and doing what the former would like to hear, no matter at what cost to the concerned unit and its various components. The junior members of the entrepreneurial corps of the industrial families need the personal support of the managers to do well and impress their seniors on the substructures and the core councils. This is suitably exploited by such managers, as are the members of the substructures and/or the confidants of the senior members of the industrial families. Sometimes, however, the equations do not work out properly and the concerned individuals from both the sides suffer considerably.

TREATMENT OF UNITS

The treatment meted out to the different economic units by the industrial families has been viewed from two angles :

(a) Operationalization of the relationship of the families with them in terms of their existence *vis-a-vis* gains from each other.

(b) The capability of the units to yield returns to the shareholders.

Operationalization of the relationship of the family with work organizations by the industrial families showed the presence of the following continuum :

$$\text{Amoral Familism} \rightarrow \text{Familism} \rightarrow \text{Fraternalism}$$

Amoral familism tends to treat the work organizations as family properties and uses them as much as possible to extract benefits for the family, even if, in the process, the objective interests of the work organizations suffer. Here the identities of the family and work organizations fuse to the extent that the latter are made the instruments for the pursuit of the goals of the former. In the case of familism, the identities of the two remain separate entities and one stands to gain from the gains of the other and not at the cost of the other. Fraternalism tends to completely merge the identities of the family and the work organizations in a positive way where the two pursue the same goals and the employees of the work organizations get benefits normally available to family members. From the fraternal evidence available on this point, three conclusions emerge (Table 8.10) :

TABLE 8.10 : **Nature of treatment of economic units by industrial families**

Nature of Treatment	Frequency and Percentage of Economic Units					
	Karta III		*Karta* II		*Karta* I	
	Frequency	Percentage	Frequency	Percentage	Frequency	Percentage
1. Amoral familism	725	57.5	441	49.0	328	61.6
2. Familism	395	31.3	341	37.9	159	29.8
3. Fraternalism	142	11.2	118	13.1	46	8.6
Total units	1262	100.0	900	100.0	533	100.0

1. The industrial families have not gone in for only one of the three choices open to them : they have tried all the three.

2. Over the years, the incidence of amoral familism has fluctuated between 61.5 per cent (first set of *kartas*), 49 per cent (second set of *kartas*) and 57.5 per cent (present set of *kartas*).

3. The occurrence of familism and fraternalism increased slightly at the time of the second set of *kartas* but has again fractionally declined at the time of the present set of *kartas*.

This study did not go into details of the comparative performance of units where the three attributes discussed earlier were present.[8] However, from discussions with the respondents it was indicated that wherever fraternalism had bestowed upon the employees the status of a partnership, the performance had been very impressive ; the work environment had been remarkably tensionfree ; and significantly, the need for the presence of the industrial bureaucracy had been greatly reduced, if not altogether eliminated.

The second aspect of treatment of the units is concerned with investment image of the industrial families in the minds of potential investors on the one hand, and existing shareholders on the other. Since the industrial families always want to carry on their manufacturing ventures with the major portion of investment coming from sources other than their own, it is essential that the image of some if not all of their economic units, measured in terms of returns to the shareholders, remains very attractive. This is expected to ensure very enthusiastic response from potential investors to their new ventures or to the expansions and diversification programmes of the existing units. Judged from this angle, the industrial families could be categorized as under :

1. Families that keep a judicious mix of the very high, high, moderate, low and no return units among the comity they control. About 52 per cent families belong to this category. These families have some very prestigious units in most of the products or product groups in which they have entered. These units show a very consistent record of profits and usually keep on issuing bonus shares every five or six years. Quite a large number of sick units are also owned or controlled by these families. Thus, even though quite a few of their units have been closed or taken over by the government, the investors respond favourably to their investment offers.

2. Families that maintain only the sound and economically very high-yielding units. This category accounts for about 15 per cent of families. These families get the most favourable response from the investors. Generally, such families have not entered into many products or product groups.

3. Families that maintain their units without loss but do not offer high returns to the investors. These account for 33 per cent of the families.

To sum up, in the management of economic units, the industrial families have shown a very partisan outlook in the issues pertaining to wage determination. In exercising dominance, they have displayed very antiquated attitudinal positions. They have penetrated the trade union structure to weaken the negotiating position of the workers. Considering the shareholders as a necessary evil, they have maximized their own returns at the cost of shareholders and reduced the latter's participation in decision making to the barest minimum and formal levels. To ensure a favourable climate for themselves, they have packed the key segments of their support systems with relations and members of their community. In treating the units they have perpetuated a system that has only their familial interests in view.

NOTES

1. A contributory factor was the declared policy of the Indian Government to keep these in the public sector in post-independence India.
2. Crouch (1977, p. 5) provides an interesting discussion on this theme.
3. Hazari (1966), Monopolies Enquiry Commission Report (1963-64) and Industrial Licensing Policy Enquiry Committee Report (S. Dutt : Chairman, 1969) have brought out the various dimensions of the control techniques. The present study does not duplicate known facts.
4. See, for example, the studies of the Reserve Bank of India (1977), the Economic and Scientific Research Foundation (1964, 1965, 1970), and the Industrial Credit and Investment Corporation of India (five series up to 1977).
5. In giving response to issues relating to the shareholders, the respondents gave their percentage estimation and not the exact numerical strength. Only percentages are consequently used in Tables 8.7 and 8.8.
6. Two important studies on this point are those of Indira Parikh (1978) and Sambrani and Garg (1978). According to Indira Parikh, the role concept of managers interferes between personality structure and social structure. The managers, therefore, accept the authority of family members, concentrate on their tasks and look for social rewards. Sambrani and Garg indicate that the industrial and civil bureaucracies are not in complete contact with the total Indian reality. As such, they become fighters of lost causes.
7. McLean and Jillson (1975, 1977) indicate that this damaging competitiveness is also present in western work organizations and is the result of need for recognition and approval, greed for power, envy, fear of criticism, and lack of consideration for others.
8. This should be a priority area of research geared to evolve more viable alternative management systems in the industrial field.

9

NATURE OF ENTREPRENEURSHIP
Negotiation of the External Environment

OPERATIONALIZATION of the external business environment by the industrial families involves negotiating favourable relationship with the state apparatus, suppliers and dealers, multi-national companies, their subsidiaries and other competitors and finally the consumers. The state decides the objectives of social transformation including industrial development and its policies pertaining to such key issues as resource allocation, capital formation, incentives, assistance, taxation, export-import trade, licensing, pricing and distribution, capacity, diversification, and entry into product or areas, etc. eventually determine the path of industrial development preferred by the country. The suppliers and dealers on the one hand, and the competitors on the other play significant roles in the evolution of their relationship with various consumer segments. The negotiation of the external environment also affects the nature of entrepreneurship displayed by the industrial families in controlling the internal environment of the industrial enterprises and, to a certain extent, even the internal functioning of the families themselves.

THE STATE

The relationship between the state and the Indian industrial families is viewed from two angles. First, how the internal organizational stresses and strains within the ruling party, that control the state apparatus, have shaped its policies and measures in respect of industrial development in general, and the role of private enterprise in it in particular. Events of this genre demarcate the operational boundaries for the private enterprise to work its way through the corridors of state apparatus. The second aspect is how within the changing political context, the private enterprise works out its relationship, specifically pertaining to its operations in the industrial field.

The British, during their long rule in India, had succeeded in making the Indian economy completely dependent on their own economic policies thus enabling them to exploit the Indian resources at a much lesser cost. The strategy to be adopted for social transformation of Indian society in the post-independence period was, by and large, decided by Gandhi and Nehru, in the pre-independence period itself. As is well known, both these leaders maintained the tactical separations between an accommodative party ideology and organization aimed at reassuring the propertied classes, and an economic strategy that incorporated proposals for institutional change. This change was designed to speed up the formation of a popular organization and to exert pressure from below on state governments for the implementation of social reforms.[1]

The First Five-Year Plan demarcated the role of private enterprise in the economy by indicating the intention of the state to increase its control and supervision over it (First Five-Year Plan, 1953 pp 31-32). The plan specified that private enterprise had to operate within the framework of the Industries (Development and Regulation) Act 1951 and that the public sector was to assume a dominant role in the industrial development of the country. By the time the Second Five-Year Plan was finalized and launched, the private sector had raised demands for major concessions on key issues of industrial policy, the essentials of which were contained in the Industrial Policy Resolution (IPR) of April 1956 (Second Five-Year Plan, 1955, pp 43-50). The IPR had classified the industries into, (i) basic and of strategic importance, (ii) essential and (iii) remaining. The first was earmarked solely for the public sector, the second was to be developed predominantly by the public sector, and the third was open to both the state and private sectors. The participation of the private sector even in the third category was subject to the regulation under the industries (Development and Regulation) Act, 1951. Thus in the envisioned socialistic pattern of society, the public sector was not only expected to grow in absolute but also in relative terms to the private sector. Failures of implementation and lack of resources, however, restricted accelerated industrial development through the public sector during the Third Five-Year Plan period. As a result, greater reliance on the private sector to stimulate growth of industrial production was reflected in a series of measures to relax controls under the Industries Act 1951.[2] At the end of Third Five-Year Plan, the private sector's contribution to the output of organized industry was still approximately 80 per cent of the total (Hazari, 1967, p. 15). Licensing systems had failed to prevent the growth of capacity in less essential industries and had, in point of fact, helped the larger business houses to usurp a disproportionate share of new licensed capacity (ILPIC, 1969, pp. 183-86). The share of the total investment allocated to the four largest houses had actually increased (Hazari, 1967, pp. 8-12).

At about this time, differences between the national leadership and the socialists inside the ruling Congress Party on the endorsement of radical policy goals and methods, surfaced more prominently immediately after the first split in 1969.[3] The Congress Forum for Socialist Action (CFSA) wanted, (i) an extensive programme of nationalization of industries in the core sector to generate vast funds required for establishing public sector plants, (ii) selective nationalization of 'non-essential' industries to capture 'lucrative heights' in cases where large profits would become available to the state, (iii) appropriate state intervention in consumer goods industries and other distribution systems and (iv) incremental nationalization of remaining parts of the economy. Techniques identified for accomplishment of the aforesaid four goals were :

(a) government participation in the management of private sector enterprises in proportion to its holding in their share capital ;

(b) conversion of loans or debentures into equity ; and

(c) pursuit of a judicious policy aimed at raising state holdings in the enterprises to reach commanding heights (Congress Forum for Socialist Action, 1971, pp. 27-31).

These demands were not met by the government in toto. However, the following steps were taken 'to curb concentration of economic power' in fewer hands :

(a) abolition of the managing agency system ;

(b) limitation of managing directorships to only one company ;

(c) approval from the Company Law Board for appointment of any future paid directors ;

(d) reduction of salaries and perquisites ;

(e) introduction of a new licensing policy following selected recommendations of the Industrial Licensing Policy Inquiry Committee 1969, which withdrew exemptions from licensing, laid specific restrictions on companies of larger industrial houses, foreign firms and their subsidiaries ;

(f) acceptance of joint sector as a concept for investment in the heavy investment sector ;

(g) directions to public financing institutions to insert convertibility clauses into loan agreements with industrial companies ;

(h) taking over of voting rights of shares pledged to public financial institutions and appointment of representatives to the board of directors ; and

(i) registration of all undertakings with the newly established Department of Company Law and satisfaction of the Monopolies and Restrictive Trade Practices Act by larger companies and foreign firms seeking expansion.

These steps further complicated the efforts towards industrial development, especially for new recruits. On the other hand, because of their intimate linkages with the effective levers of power in the government, the large business houses short-circuited this regulatory system with aplomb. By taking these steps, the government also avoided nationalization, and tended to help concentration by directing larger houses to invest in massive projects. These measures have not been scrapped since their initiation, although by now a familiar pattern of effectively circumventing them is very well known. After every two or three years, the economy gets into difficulties, displaying high inflation, declining or slow growth of output, lack of resources, etc. The class collectives of the industrial families, the Federation of Indian Chambers of Commerce and Industries (FICCI) and the Associated Chamber of Commerce and Industry (ASSOCHAM) raise a hue and cry before the government, complaining of unhelpful investment climate and other difficulties. The ruling party government responds by ordering concessions on most of the regulatory provisions.[4]

The second major difference between national leadership and the Congress Forum for Socialist Action was pertaining to constitutional change to get over the obstacles to economic and social reforms contained in the existing provisions. The Government of India did amend the Constitution several times, but matters came to head with the 17th Amendment (1964) primarily enacted for the speedy implementation of land redistribution. The Supreme Court repudiated it in 1967 in the famous Golak Nath case. The private sector and the propertied classes continued to benefit through prolonged legal battles with the government. Another round of constitutional amendments between 1974 and 1977, enabled the government to arm itself by initiating some changes in the structure of social relations in Indian society. The impasse continues to this day, since the state is not committed to any radical change in the social structure without which the 'stated' social and economic goals of planning exercise initiated during the last 33 years remain unrealized.

The following emerging principles from Hindu custom and law which govern ownership, succession, inheritance and taxation of the dominant majority of Indian industrial families have also influenced industrial development in a very decisive way :

1. Corpus of joint family : The history of a Hindu family, as unfolded by the *Sutras*, *Smritis* and other religious texts, indicates that even though individuals have been progressively granted greater rights, it is also ensured, at the same time, that these rights harmonize with the interests of the family. Thus in theory the individual can, for example, demand his share from the family property. However, the prevelent custom ensures the continuance of the corpus of the family since no incentives are given to a person to breakaway from the family.

2. Recognition of Difference of Joint Family in Custom and Law : Hindu Law makes a distinction between the coparcenary and the joint family. According to Kane (1946, p. 591), under the *Mitakshara* law,[5] a Hindu coparcenary is a much narrower group than the joint family. It comprises only those males who take an interest in the joint or coparcenary property

by birth. Under the *Dayabhaga* law, there is no coparcenary between a man and his sons during his life time, even though they may be living in a single household. The important point to be noted, however, is that *Hindu law* (under both the schools) *is not concerned at all with the distinction between the elementary family and the joint or extended family.* It is concerned primarily with the *constitution of the property-holding group and with the persons having rights of maintenance from the property-holding group, and not with the constitution of the household group* (Shah, 1973, pp. 117-18) emphasis added.

3. Fluidity and flexibility of Legal form of family : This principle flows from the peculiar law and practice governing succession to joint family property (Gulati and Gulati, 1962, p. 1). Under *Mitakshara* law, each son acquires at his birth an equal interest with his father in the joint family property and on the death of the father the son takes the property, not as his heir, but by survivorship (Mulla, 1959, p. 47). Under the *Dayabhaga* law, the son's rights arise for the first time on the father's death to the extent that the property is left by the father as the father's heir. In other words, it does not recognize the principle of survivorship (Mulla, 1959, pp. 314-15, 417-19, 421). This being so, every member of the *Mitakshara* family has only an undivided interest in the joint family property. Under the *Dayabhaga* law, a member of the family holds his share in quasi-severally and it passes on his death to his sons.[6]

The extent of share a coparcener has in the family property is defined under the *Dayabhaga* system, it is not defined in *Mitakshara*. The family property ordinarily is managed by the *karta* who enjoys a wide range of powers. He is supposed to use the income from the property for the purposes of the family. If other coparceners are not satisfied with the manner in which the *karta* exercises his powers, their own remedy is to demand a partition (Gulati and Gulati, 1962, pp. 3-4). The essence of a *Mitakshara* joint family is the unity of ownership : the partition consists in ascertaining and defining the shares of coparceners (Mulla, 1959, pp. 492-93). Partition under *Dhayabhaga* is easier, since the shares are already known. A partition between the coparceners may be either complete or partial. This partial partition may be in respect of persons or property. Part of the joint property may be partitioned and the residual kept as joint (and undivided) property. Non-separating coparceners can remain joint and enjoy what remains of the joint property. A reunion of separated coparceners is also recognized (Mulla, 1959, pp. 503-08).

Treatment of Hindu Undivided Families (HUF) as a separate entity for tax purposes and the facility of partial partition and the option to rejoin the deserted coparcenery group, confer tremendous tax benefits. Gulati and Gulati (1962) have shown that it helps the avoidance of tax in a number of ways.[7] On the surface it seems that the contradiction between the objectives of social customs and the enforced Hindu Law are unmanageable. In reality it is not so. The industrial families extract maximum benefits using both the societal custom and enforcement of law.

NEGOTIATION OF RELATIONSHIP WITH STATE BY INDUSTRIAL FAMILIES

Within the ambit of the existing industrial climate, the industrial families have negotiated their relationship with the state. This process of negotiation has been examined from three angles by Kochanek (1974), Venkatasubbiah (1977) and Erdman (1973). Kochanek treats the industrial families as interest groups which organize themselves for collective action and which attempt to influence the policies and decisions of the Government of India. According to him (1974, p. 52), interest groups respond and adapt to the structures of power in any given political system. Changes in the structure of government, or changes in the distribution of power within the structure of the government necessitate changes in the various sections of society which accordingly realign their channels of influence. Venkatasubbiah discusses the problems between the government and organised business and indicates how the business community in its dealing with the government, itself differs or agrees. He postulates that the unity of the industrial families as a class is valid only to the extent that it promotes their own interests and therefore, he concentrates on bringing out sharply the differences that exist among them in dealing with various issues. Erdman's study deals with the teaming up of private capital and management skills of industrial families with the finance and 'control' of the government, a situation in which the performance of entrepreneurial functions by the management set-up involves the interplay of the government's business and regulatory profiles. The present study benefits from each of them. However, it does not discuss the strategies themselves as such, but rather their benefits reaped by the industrial families on various issues. These issues are :

(a) general alignment with the leading lines of relationship with the government[8] ;

(b) the extent of use of the police apparatus of the states ;

(c) the nature of state intervention obtained in disputes involving industrial families ;

(d) the ability to exploit lacunae in the industrial laws ;

(e) payment of taxes and dues to the state, and other autonomous agencies ;

(f) the use of regulatory mechanisms (licensing, etc.) ;

(g) the exploitation from industrial sickness ;

(h) allocations of funds from industrial financial institutions and banks ;

(i) the extent of influence exercised on the fixation of prices and other fiscal and economic measures ; and

(j) the appointment or continuation of appointees in key economic positions in government and financial institutions.

On each of the above aspects, the industrial families have two channels of access to the government : (i) through their collective organizations such as FICCI/ASSOCHAM (or numerous such regional and industrial product associations) and (ii) their liaison substructures at family levels. The latter is generally manned by retired, high government officials and professionals, who see that the files move, provide access to key officials, and get the work done. The last two aspects, however, are more open to exploitation by the industrial families as a class, although in the exertion of such influence some among them play a more prominent part.

It is essential to emphasize the nature of the data used in the analysis that follows. The data have been obtained from the respondents and their associates, and are opinions regarding various aspects of the relationship of the industrial families with the state apparatus. Although in terms of their utility they are of great significance, the fact remains that they are not the actual facts themselves.[9]

The Birla line of establishing a going relationship with the state has continued to enjoy overwhelming support from the industrial families in preference to the Tata line (Table 9.1). Generally, this indicates the basic character of the industrial families as a class. This majority does not want to condemn black sheep among them and, in general, wants to appear politically neutral. From a purely political point of view, this neutrality means nothing since the industrial families cannot possibly afford to annoy the sovereign state at any point of time. Therefore, in effect it amounts to toeing the line of the ruling party. By the same token it also means no open support to the opposition parties. This approach is not something to be proud of. Nevertheless, it is the shrewdest possible stance the industrial families can adopt in extracting maximum possible benefits for the minimum given in return. In the actual exchange of the *quid pro quo* it is persuasive in dealing with the selected few from the ruling party.

Inside the economic units, the industrial families and their support systems do not exactly enjoy the confidence of the workers and other employees, the latter constituting the exploited segment in these organizations. Quite often in handling disputes involving workers and their trade unions, the industrial families require the use of the police apparatus of the state. In the eyes of law, of course, the police are supposed to be protectors. In practice, however, they generally end up on the side of the powerful irrespective of the facts of the particular cases. The use of the police by the industrial families is not for the protection of the workers : almost always it is to control and suppress their legitimate struggles. None of the industrial families have denied the use of the police for their purposes. While the moderate use of the police force has remained stabilized at 39 per cent, high police use has somewhat appreciated (from 30.3 per cent to 33.3 per cent to 36.4 per cent) and low police use has declined (from 30.3 per cent to 27.3 per cent to 24.2 per cent). Directionally, therefore, larger number of industrial families are tending to use police reinforcements.

The nature of the existing machinery to settle labour disputes is such that the state apparatus becomes a mediator in all major cases and certainly in those involving protracted disputes resulting in a stalemate. Even in the reference of cases to the labour courts or tribunals, the state plays a key role. In

TABLE 9.1 : Nature of relationship with the state

Aspects and Nature of Relationship	Frequency and Percentage of Families					
	Karta III		*Karta* II		*Karta* I	
	Frequency	Percentage	Frequency	Percentage	Frequency	Percentage
1. *Relationship with the Government*						
a. Birla line	20	60.6	22	66.7	24	72.7
b. Tata line	13	39.4	11	33.3	9	27.3
2. *Use of Police Apparatus of the State*						
a. High	12	36.4	11	33.3	10	30.3
b. Moderate	13	39.4	13	39.4	13	39.4
c. Low	8	24.2	9	27.3	10	30.3
3. *Nature of State Intervention*						
a. Partisan	23	69.7	16	48.5	13	39.4
b. Evenhanded	10	30.3	17	51.5	10	60.6
4. *Exploitation of Legal Lacunae*						
a. High	23	69.7	20	60.7	18	54.6
b. Moderate	6	18.2	8	24.2	8	24.2
c. Low	4	12.1	5	15.1	7	21.2
5. *Outstanding Taxes and Dues*						
a. High	16	49.5	13	39.4	9	27.3
b. Moderate	12	36.4	12	36.4	14	42.4
c. Low	5	15.1	8	24.2	10	30.3
6. *Use of Regulatory Mechanism*						
a. Disproportionate	18	54.5	13	39.4	9	27.3
b. Adequate	9	27.3	13	39.4	16	48.5
c. Inadequate	6	18.2	7	21.2	8	24.2

7. *Exploitation of Sickness*						
a. High	10	30.3	7	21.2	4	12.1
b. Moderate	9	27.3	8	24.2	9	27.1
c. Low	8	24.2	10	30.3	13	39.4
d. None	6	18.2	8	24.2	7	21.2
8. *Allotment of Funds*						
a. Disproportionate	15	45.5	12	36.4	9	27.3
b. Adequate	10	30.3	14	42.4	16	48.5
c. Inadequate	8	24.2	7	21.2	8	24.2
9. *Influence of Fiscal and Economic Measures*						
a. Considerable	12	36.4	9	27.3	6	18.2
b. Moderate	15	45.5	14	42.4	14	42.4
c. Low	6	18.2	10	30.3	13	39.4
10. *Appointments/continuation of Key Economic Positions*						
a. High	9	27.3	8	24.2	3	9.1
b. Moderate	17	51.5	16	48.5	16	48.5
c. Low	7	21.2	9	27.3	14	42.4

the tripartite machinery, of course, it constitutes the third party, the other two being the workers represented through their trade unions and the managements of the units concerned. Since the industrial families maintain intimate relations with key segments of the ruling parties and the latter are equally involved in trade union politics, state interventions in labour disputes become exercises not always above partisanship and favouratism. Opinionistic evidence from this study indicates that over the years the nature of state interventions has shown an alarming tendency of being partisan, favouring the industrial families. Partisan interventions at the time of first set of *kartas* were about 39 per cent : at the time of present set of *kartas* this percentage has gone up to 69. Two very significant developments during the last few years have been a disquieting increase in the number of prolonged lockouts and the abject failure of almost all prolonged, total strikes. At the end of both types of these developments, it is only the workers who have suffered. This is perhaps convincing indication of the nature of state interventions and illustrates the unequal nature of the abilities if employers and employees to influence the state apparatus.

The collective organizations of the industrial families (FICCI, ASSOCHAM etc.,) ensure through their liaison network that the industrial legislation enacted by Parliament contains enough loopholes. These are exploited later on by the families singly or, if necessary, by a number of them through concerted legal action. Their monetary power equips all of them with tremendous capacities to fight legal battles. As a result, they are capable of denying, for protracted periods, the fruits of legislation to the workers. However, use of this option has largely depended upon the family traditions, and immediate and future gains accruing to them. A majority of industrial families—54.6 per cent to begin with, and 60.7 and 69.7 per cent at the time of the second and third sets of *kartas* have highly exploited the legal lacunae left in industrial legislations (Table 9.1).

The industrial undertakings as corporate bodies and the industrial families as collective entities (and family members as members of Hindu undivided family, family trust, private trust, etc.) owe tax obligations to the state and other agencies. For the industrial undertakings it includes a wide variety of taxes and cess. It also includes levies to the state and payments for the use of services or inputs to such agencies as civic bodies, electricity boards, etc. In addition, there are collections of provident fund, income tax, etc. to be made and submitted to the required agencies. At the individual and family levels these include income, wealth and gift taxes super. The industrial families as a class believe in running their business and industry with capital supplied by someone else but preferably an anonymous body such as the state. Their histories of growth in business, trade and industry are, among other things, equally stories of accumulating huge arrears of unpaid taxes and other dues. Empirically, the evidence from this study and regular pronouncements in the budget sessions of Parliament each year suggest that this practice has greatly increased among all industrial families over the years. This is evident from the increase in the percentage of industrial families admitting huge tax arrears and decline in percentage of those families with low tax arrears.

The State in India, having opted for what it termed 'mixed economy' had earmarked certain areas for the public sector to reach commanding heights and left the others in the hands of private capital. The operation of private capital has, however, been subject to regulation of licensing and other allied measures to curb concentration and unfair trade practices. After the abolition of the managing agency system in the mid-sixties, Parliament enacted the Monopolies and Restrictive Trade Practices (MRTP) Act. Of course, licensing system of industrial units also continues. The working of the total regulatory system was aimed at preventing unfair advantages to a few families. However, between the tenures of the first and third sets of *kartas* the percentage of the industrial families deriving disproportionate benefits in spite of the existence of regulatory mechanisms has jumped from 27.3 to 54.4.[10] The devices being used to obtain disproportionate benefits from the regulatory system include : excess capacity installation, its sly regularization, capacity preemption, diversion to floated small units, and the like.

The analysis of industrial sickness in private sector enterprises in India has often been value-loaded and partisan. The workers and their trade unions feel that the managements bring it about through their systematized misdeeds. The managements on the other hand attribute it to labour unrest, heavy taxation, unfavourable markets, entry of alternate products, and technological obsolescene. The state, as a mediator, takes a position that includes all these factors in different degrees in different contexts. Apart from the worker's point of view, industrial sickness has generally been considered to be an outcome that is value-free and without any conscious motive, intent and design : Because of industrial sickness, it is stated, those who own and control the enterprises always stand to suffer and lose. Discussions with knowledgeable persons in this study, however, reveal that :

1. The Indian industrial families themselves continue to grow, although quite a few of their units become sick, and some even get closed.

2. Sickness does not appear all of sudden just from no where.

3. Chronic symptoms of sickness appear much later whereas the root causes quite often lie in the decisions taken or not taken much earlier.

4. Sickness occurred quite often in such units in which the personal financial stake of the controlling industrial families was negligible and hence it was more profitable to go in for a conscious policy of milking such units bone-dry than making further investments in them.

5. The industrial families quite often bring about sickness in such a planned manner that they stand to gain rather than lose in the process.[11]

With this background in view, the matter was further investigated with the help of all segments of the manufacturing process. Evidence offered by them indicated that the tendency to exploit industrial sickness has been present in the majority of industrial families all through, only its extent has varied. Over the years, the percentage of industrial families admitting no exploitation has more or less remained constant around 20 per cent ; that showing high exploitation has increased from a low of 12.1 per cent at the time of the first set of *kartas* to 21.2 per cent and 30.3 per cent, respectively, at the time of the second and third sets of *kartas*.

The Industrial Licensing Policy Inquiry Committee (1969) and Chowdhuri (1975) have brought out the discriminatory nature of the allotment of funds to various industrial families by public financial institutions. Evidence from the present study also confirms this trend. It is evident that nationalization of banks and directives given by the government to public financial institutions to avoid the use of funds primarily for the large industrial families have not been very successful in reversing the trend.

Fixation of prices, decisions about other fiscal and economic measures and appointments, continuation of appointees in key economic positions (i.e. the Governor of Reserve Bank of India, the chairmen and members of the boards of financial institutions and banks, the ministers and secretaries of the union ministries of Finance, Industries and Labour, etc.) are some spheres where the industrial families operate as a class and exert their influence through their collective entities. Opinionistic data from this study suggests that this type of influence, which by its very nature cannot be open, is still quite considerable. What is more remarkable is the fact that in its intensity it has tended to increase with the passage of time. Overall, the industrial families have shown an extraordinary capacity to control the state apparatus in their quest for maximization of profits.

INPUT SUPPLIERS AND OUTPUT DEALERS

Without being part of the internal system of the factory, input suppliers and output dealers, through their linkages with the inputs and outputs (and consequently through-puts), vitally affect their internal functioning. There has to be, therefore, a special category of people to be selected for these assignments.

From the response of the respondent families it appears that there are three such major categories, (i) members of kin groups, (ii) members of the same (i.e. trade, business) community and (iii) others (Table 9.2). The strength of these groups varies over the time continuum. The average strength of the input suppliers was 55.3 (first set of *kartas*), 85.9 (second set of *kartas*), and 122.4 (third set of *kartas*). The average strength of the output dealers, on the other hand, was 177.7, 343.6 and 734.2 at the same points of time. Between input suppliers and output dealers, the strength of the latter was appreciably larger.

Strong reliance on the kin groups and community for input supply and output disposal is displayed when the composition of the two groups is examined. Between input suppliers and

TABLE 9.2 : **Composition and strength of input suppliers and output dealers**

Composition of	Frequency, Average and Percentages								
	Karta III			*Karta* II			*Karta* I		
	Frequency	Average	Percentage	Frequency	Average	Percentage	Frequency	Average	Percentage
				Input Suppliers					
1. Kin group	1855	56.2	45.9	1268	38.4	44.7	710	21.5	38.9
2. Community members	1406	42.6	34.8	1234	37.4	43.5	834	25.3	45.7
3. Others	778	23.6	19.3	333	10.1	11.8	281	8.5	15.4
Sub-total	4039	122.4	100.0	2835	85.9	100.0	1825	55.3	100.0
				Output Dealers					
1. Kin group	13213	400.4	54.6	5293	160.4	46.4	2894	87.7	49.3
2. Community members	6894	208.9	28.4	3221	97.6	28.4	1818	55.1	31.0
3. Others	4122	124.9	17.0	2825	85.6	24.9	1152	34.9	19.7
Sub-total	24229	734.2	100.0	11339	343.6	100.0	5864	177.7	100.0
				Input and Output Dealers					
1. Kin group	15068	456.7	53.3	6561	198.8	46.3	3604	109.2	46.9
2. Community members	8300	251.5	29.4	4455	135.0	31.4	2652	80.4	34.5
3. Others	4900	148.5	17.3	3158	95.7	22.3	1433	43.4	18.6
Grand total	28268	856.7	100.0	14174	429.5	100.0	7689	233.0	100.0

output dealers, higher percentages of kin groups are consistently present among the latter (38.9 per cent, 44.7 per cent and 45.9 per cent against 49.3 per cent, 46.3 per cent and 54.6 per cent respectively). In both cases, the percentage of the kin group has continued to increase. The proportion of members of the same community is equally high among input suppliers and output dealers. However, it has tended to decline marginally. The percentage of 'others' seldom exceeds 25.

The usual explanation given by the respondent families for a low percentage of 'others' was two-fold, one that they lacked the necessary capital and two that, they did not possess the requisite qualifications (i.e. contacts, business skills, accounting ability, etc.). However, it was only when the terms of agreements, interlinkages of input suppliers and output dealers and their composition are closely examined that the real motive behind this arrangement becomes clear. It becomes known, for example, that quite a few of these firms are ones in which the industrial families themselves are partners. To that extent they stand to gain even in such situations where the manufacturing unit (to which they are supplying raw material or from which they are lifting finished products) is losing. The firms in which the industrial families have themselves a stake are given unusually attractive terms and have very little to do by way of services to be rendered. It is also revealed that some of these firms just do not exist in reality; this state of affairs is not exactly unknown to the government as evidenced by the existence of the Companies (Appointment of Sole Agents) Rules of 1975.[12] Thus, by using the kin groups and community in these mechanisms the industrial families draw support from traditional collectivities in their economic activities to fortify their positions, minimize their losses and maximize their gains.

CONSUMERS

Consumers form the last component of the manufacturing chain that starts with the acquisition of raw materials and ends with the sale of finished products. In terms of importance they rank next only to the state among all components of the business environment. There can be occasions where they can upset the efforts of the businessmen and the state in exploiting them by simply refusing to buy the products. The consumers, however, can exert their influence on the industrial families only when they assert their will : If they fail to do so, as is the case in India because of historical reasons,[13] their influence can be but minimal.

Corporate and product development cycles decide the consumer management strategies adopted by the entrepreneurs.[14] The industrial families in India have used the following consumer management strategies.

1. Products : In choosing the products to manufacture, they have been guided by their marketability, capital investment, gestation periods and development of new technology. The ones they have preferred most are mass consumption items. In the post-independence period, some products were kept out of bounds for them as a matter of state policy. But they have themselves not entered into quite a few of their own free choice. In fact the corporate profiles of each of the industrial families indicate that they entered one particular product, gained eminence, consolidated, and then diversified into other lucrative product groups.

2. Production Technologies : By and large, the industrial families went for proven technologies and did not themselves develop any major technological processes. Whatever technological choices were made during the post-independence period, were obtained from the western hemisphere and, in a few cases, from Japan. These collaborations did not enhance their research and development capabilities.

3. Quality of Products : The segmentation of consumers has been attempted in an organized manner. This enables the manufacturers to create a so-called 'class-consciousness' among a certain type of consumers to buy some prestigious products. The quality and design of these products is, of course, impressive, but is in no way commensurate with the prices charged. The other extreme in market segmentation is provided by the consumers who are very poor. They want low quality products. The volume of sales in this category is very large, although single unit margins are decidedly not fabulous. In totality, margins are fair. Some among the industrial families have concentrated on the higher class consumers whereas certain others have opted for the poor.

4. Quantity of Products : Most mass consumption items quite often remain in short supply. As a matter of policy, their

production capacities have not increased to cope with the increasing demand. Scarcity of items enable output dealers to work out, at their level, preferential segments among consumers. In recent times, when capacities of some such items have reached a stage where demand and supply are equal or where the latter has slightly outstripped the former, the manufacturers have formed cartels and have resorted to production cuts or restricted releases to bring back scarcities.

5. Pricing and Marketing : Three broadbased strategies are in use : competitive pricing (for such items as are available in ample measure from alternative sources) ; low price and high volume (for mass consumption items among the poorer segments of the population) ; and high price items (for well-to-do segments of the population). Marketing, selling, distributing channels and segments are determined by the status of the product in the market ; the financial condition of the company or family, and the consumer segment to which they are addressed. The higher the status, the more favourably loaded are the conditions for the families.

Another dimension of pricing and marketing is what the MRTP Act describes as unfair trade practices. A perusal of the action initiated by the MRTP Commission and the verdicts given indicate that a majority of the respondent families figure in them[15]. However, some among them have committed a larger number of such offences. The alliance between the industrial families and multinationals plays a significant role in cartel formation and escalation of prices—devices by which ultimately the consumers suffer.

MULTINATIONALS AND OTHER COMPETITORS

Industrial development in India has been dovetailed with the exercise of sovereign political power by the British during the pre-independence period, and by the ruling political parties (the Congress for over three decades, the Janata between March 1977 and July 1979, the Janata[s]-Congress coalition between July and December 1979, and Congress [Indira] thereafter) after independence. In this process, foreign and Indian industrialists have played different kinds of roles during various phases. During the British period, the Indian economy was completely dependent upon the British, and the British and Indian businessmen constituted two opposite camps. Since Indian businessmen were subjects and foreign companies were associated with the imperial power, the permitted role for Indians during this period, was definitely inferior in status and scope.[16] Some among the Indians, especially the Parsees, did bend their backs to perform comprador roles ; others, who were not so 'loyal' to the British, still had to work within the limits of the business environment created for them by the latter.

After independence, the transfer of political power necessitated a change in this dependency relationship. Multinationals now had to seek political support from the Indian government. In the economic transition that took place, quite a few foreign companies in India passed into Indian hands through the managing agency system. As Indian businessmen enlarged their manufacturing operations, a realignment of their relationship with the multinationals took place. Multinationals and Indian businessmen had become friends. In some cases, with the transfer of interests of some foreign companies to Indians collaboration arrangements with the parent companies have continued. In the changed political climate it became easier for multinationals to exploit as large a market as India with the help of Indian industrialists, since it was not possible for them to do so directly. Over 80 per cent of the industrial families have collaboration agreements with multinationals. Beginning in the early sixties, the world capitalists economy itself has been undergoing structural changes which have forced the multinationals to undertake a worldwide reorganization of their production through relocation of production to new sites (in particular in developing countries), as well as through speeding up the rationalization of production at the traditional industrial sites.[17] A detailed study would probably indicate the vital link provided by the Indian industrial families in this schema.

Internal differentiation within the industrial families was always present. During the British period this was, at its highest level, reflected in the comprador and nationalist profiles and in the formation and functioning of the Federation of Indian Chamber of Commerce and Industry (FICCI) and the Associated Chamber of Commerce and Industry (ASSOCHAM). After independence this differentiation was best illustrated by the differences among them over condemnation of those using unfair practices, formation of the Fair Trade Practices Association and such consumer-oriented issues on the one hand, and floating of the Swatantra Party and use of the state apparatus for building industrial empires on the other.

By and large this differentiation does break into easily identifiable ethnic lines, indicating the crucial role of traditional collectivities in class consolidation and intra-class conflict. The Marwaris, for example, consider the Parsees and the Gujaratis to be less than equals ; the latter consider others such as the Maharashtrians, Bengalis, Punjabis and Chettiars as some kind of lower castes among the industrialists. Yet they all collaborate with each other on issues that affect them all as a capitalist class. This co-oparation is seen in :

(a) their lobbying with the government on general economic and fiscal policies affecting them as an aggregated class ;

(b) acting as 'sleeping' directors on joint-stock companies controlled by others ;

(c) effecting production cuts, price increases, etc. in a concerted manner to beat consumer interests ;

(d) the whole gamut of issues determining the exchange relationship with the workforce as a conscious device to keep the bargaining position of the latter at a lower web ;

(e) forging a link with multinationals in an even larger effort directed at exploiting the internal and external markets ; and

(f) cementing their close business relationships by entering into corporate marriages.

To conclude, in negotiating the external business environment, the Indian industrial families exhibit a considerable capacity to control and influence the state apparatus, pack input suppliers and output dealers with relations and members of their community, tie up with multinationals and their Indian competitors to give a raw deal to consumers.

NOTES

1. Frankel (1978) provides a very vivid, well-documented and fairly up-to-date account of the goals of planning, inherent contradictions in them, and the impasse in which its implementation has landed the Indian economy.

2. For example, forty-two industries subject to licensing were delicensed ; existing units allowed expansions/diversifications ; and controls on prices and distribution of several industries were lifted. This was to become standard technique in the years to follow.
3. Such differences had existed even before the first split. Sardar Patel, Morarji Desai and S.K. Patil representing the propertied classes had obstructed the socialistic thrust of many programmes, through their dilutory actions from their safe and powerful positions in the government.
4. One of the major motivations for not scrapping these altogether is the periodical need of the ruling party to collect party funds for election and other purposes ; threats of sanctions prove useful in 'persuading' the unwilling ones to contribute.
5. Hindu Law, as is well known, has two schools : *Mitakshara* and *Dayabhaga*. The former is dominant one as it applies to most parts of the country excepting Bengal where the latter one is applicable.
6. As Shah (1973, p. 114) points out patrilineage is the only relevant group in most parts of India. However, a few matrilineal groups do exist in several parts of the country.
7. In particular, the exemption limits for the Hindu Undivided Families are higher ; there is higher exemption on the surcharge. Financial jugglery about partial partition (several varieties of it) can reduce, or altogether avoid taxes ; and transfer of assets to wife or wives of coparceners and children enables separate exemptions from paying tax.
8. These include the Birla and the Tata lines. The Birla line advocates : 1. Never challenge the government directly/openly. 2. Play along with government, step on their shoes but not too hard. 3. Work out agreements with government officials in private. 4. Never condemn the business class. The Tata line suggests : 1. The business class should be open and politically more active. 2. Openly espouse the virtues of free enterprise, and 3. Condemn the anti-social elements in the business community. Ultimately, the differences between the two lines of actions are reflected in the following steps taken by the followers of the Tata line : (a) dissociation from the FICCI which was and continued to be dominated by the Birlas and (b) formation of the Swatantra Party, Forum of Free Enterprises and Fair Trade Practices Association.
9. It is not that the 'facts' themselves are absolutely free from controversy or are the last word on the subject. The amounts due as taxes from the industrial families, for example, have been used as 'facts' by many agencies. However, the families call them "exaggerated claim" and dispute them in the law court.
10. Khurana's (1979) study evaluating the impact of MRTP corroborates this finding.
11. Ghosh (1979) very graphically indicates the route and rationale of exploitation of industrial sickness by an industrial family.

12. Through an amendment effected in these rules on 22 January 1979, the Government of India tried to plug a few loopholes.
13. Three among them are crucial, (i) the uneven nature of industrial development. (ii) characteristics of the consumers themselves and (iii) the character of the state. A detailed description of these is provided by Bagchi (1972), Chowdhuri (1975), Pavlov (1964) and Verma (1979).
14. According to Bruce (1976, p. 28), product development cycle involves the following stages, (i) research and development, (ii) marketing development, (iii) growth, (iv) saturation and (v) decline. In operationalizing these stages, dealing with product characteristics, quality and quantity produced, and pricing, the entrepreneurs decide their response to the consumers. He categorizes consumer management strategies into innovative, imitative and derivative.
15. This is also the finding of Khurana (1979).
16. Birla (1966) provides a very revealing description of the same.
17. This is the conclusion reached by Frobel et al. (1977).

10

FAMILY AND MANAGEMENT OF ENTERPRISES

Empirical Testing of Hypothesis

As indicated earlier (Chapter 4), the analysis of empirical data collected for this study is a two-stage exercise. In the first stage, consisting of Chapters 4 to 9, the demographic and authority structures of the industrial families, the nature and extent of turbulence, the strategies employed to combat turbulence, the development and deployment of financial resources, the control of the internal environment of the economic enterprises, and the negotiation of the external business environment by the industrial families have been considered in isolation. In this chapter, the interrelationships between the different dependent and independent variables are attempted as the second stage of data analysis. As a part of this exercise, four main and three related hypotheses are empirically tested. Two of these main hypotheses deal with the relationship of various types of demographic changes with the changes in the authority and management structures. The remaining two are concerned with the nature of the authority structure and the presence of turbulence and business performance with the effectiveness of strategies employed to combat familial turbulence and control of internal and external business environments. Statistical validation of the hypothesis has been made with the help of correlation coefficients.

FIRST HYPOTHESIS

Changes in the demographic structures of the industrial families are likely to change the authority structures of the families.

On empirical testing, however, it is found that of the five types of demographic changes (i.e. births, deaths and retirements, adulthood reached, marriages, all demographic changes) occurring in the industrial families, none has significantly and consistently changed the *nature* of the authority structures (Table 10.1). On

TABLE 10.1 : **Correlation coefficients between demographic changes and various dimensions of nature of authority structure.**

Demographic Change Variables	Coefficients *Karta* III	*Karta* II	*Karta* I
	Attention Control		
1. Births	.022	0.0	—.158
2. Deaths and retirement	—.046	0.0	.092
3. Adulthood reached	.043	0.0	.044
4. Marriages	.077	0.0	.025
5. All demographic changes	.051	0.0	—.022
	Person Control		
1. Births	—.020	—.361**	—.158
2. Deaths and retirement	—.221	—.318	—.091
3. Adulthood reached	—.187	—.208	.044
4. Marriages	—.074	—.119	.025
5. All demographic changes	—.145	—.255	—.022
	Attributed Authority in Domestic Spheres		
1. Births	.168	—.149	—.330
2. Deaths and retirement	.048	—.199	—.002
3. Adulthood reached	—.051	—.220	—.050
4. Marriages	—.199	—.108	—.224
5. All demographic changes	—.029	—.176	—.225
	Attributed Authority in Economic Sphere		
1. Births	—.109	—.180	.108
2. Deaths and retirement	.191	—.143	.254
3. Adulthood reached	.106	.063	.083
4. Marriages	.305	.142	.137
5. All demographic changes	.165	.023	.149
	Aggregated Authority Structure Index		
1. Births	.018	—.298	—.177
2. Deaths and retirement	—.024	—.283	.127
3. Adulthood reached	—.038	—.159	.032
4. Marriages	.057	—.042	—.023
5. All demographic changes	.006	—.179	—.048

** Significant at 5 per cent level of confidence.

the other hand, the *composition* of the authority structure has changed significantly owing to births (at the time of the first set of *kartas*), deaths and retirements (at the time of the first and second sets of *kartas*), adulthood (at the time of the present set of *kartas*), marriages (at the time of the first set of *kartas*), and all demographic changes taken together throughout (Table 10.2). Barring one case in which there is no relationship between deaths

TABLE 10.2 : **Correlation coefficients between demographic changes and changes in the composition of authority structure**

Demographic Change Variables	Coefficients *Karta* III	*Karta* II	*Karta* I
1. Births	.188	.333	.504*
2. Deaths and retirement	0.0	.529*	.417*
3. Adulthood reached	.410*	.184	.287
4. Marriages	.318	.262	.424*
5. All demographic changes	.373**	.344**	.494*

* Significant at 1 per cent level of confidence.
** Significant at 5 per cent level of confidence.

and retirements and changes in the composition of the authority structures at the time of the third set of *kartas*, in all other instances the coefficients are positive although statistically insignificant.

To find out a possible explanation for the absence of a significant relationship between demographic changes and the nature of the authority structures in the industrial families, changes in the composition and nature of authority structures were correlated (Table 10.3). It emerges that there is no significant relationship between the two sets. The first main hypothesis is consequently only partially validated.

SECOND HYPOTHESIS

Changes in the demographic structures are likely to change the deployment of family members in the management structures of the economic enterprises

Empirically, all demographic changes taken together are positively and significantly correlated with the changes in the investment,

TABLE 10.3 : **Correlation coefficients between changes in the composition and the nature of authority structure**

Nature of Authority Structure Variables	Coefficients		
	Karta III	*Karta* II	*Karta* I
1. Attention control	.354**	0.0	—.120
2. Person control	.005	—.282	—.120
3. Attributed authority in economic sphere	—.116	—.097	—.280
4. Attributed authority in economicsphere	.195	.097	—.232
5. Aggregated authority structure index	.123	—.132	—.100

** Significant at 5 per cent level of confidence.

income or expenditure structures, super substructures and core councils at the family levels, and formal management structures at the unit levels (Table 10.4). They are also positively and

TABLE 10.4 : **Correlation coefficients between all demographic changes and changes in management structures**

Family Management Structure Variables	Coefficients		
	Karta III	*Karta* II	*Karta* I
1. Number of new units	.413*	.699*	.481*
2. Number of major expansions/ diversifications	.415*	.628*	.592*
3. Number of changes in investment structures	.924*	.925*	.898*
4. Number of changes in income/ expenditure structures	.937*	.955*	.592*
5. Number of changes in the deployment of family members in unit managements	.530*	.783*	.560*
6. Number of changes in core councils	.372**	.344**	.494*
7. Number of changes in super substructures	.595*	.692*	.514*

* Significant at 1 per cent level of confidence.
** Significant at 5 per cent level of confidence.

significantly correlated with the number of new economic enterprises established and the number of major expansions or diversifications carried out by the industrial families.

Births show a positive and statistically significant relationship with the changes in all the management structures (except with the number of changes in the core councils) considered at the time of the first and second sets of *kartas* (Table 10.5). At the time of the third set of *kartas*, however, the impact of births seems to have considerably waned, with the result that they show a significant

TABLE 10.5 : **Correlation coefficients between births and changes in management structures**

Family Management Structure Variables	Coefficients		
	Karta III	*Karta* II	*Karta* I
1. Number of new units	.065	.462*	.400**
2. Number of major expansions/ diversifications	.038	.443*	.372**
3. Number of changes in investment structures	.624*	.694*	.616*
4. Number of changes in income/ expenditure structures	.828*	.868*	.372**
5. Number of changes in the deployment of family members in unit managements	.143	.512*	.439*
6. Number of changes in core councils	.188	.333	.504*
7. Number of changes in super substructures	.315	.651*	.606*

* Significant at 1 per cent level of confidence.
** Significant at 5 per cent level of confidence.

relationship only with changes in the investment and income or expenditure structures. In all other cases the coefficients are positive but statistically insignificant.

Deaths and retirements show a very fluctuating relationship with the changes in the management structures at different points of time (Table 10.6). During the tenures of the third set of *kartas*, they are significantly correlated only with changes in the investment and income or expenditure structures. The situa-

TABLE 10.6 : **Correlation coefficients between deaths and retirement and changes in management structures**

Family Management Structure Variables	Coefficients		
	Karta III	*Karta* II	*Karta* I
1. Number of new units	.267	.347**	.292
2. Number of major expansions/ diversifications	.308	.507*	.181
3. Number of changes in investment structures	.606*	.451*	.470*
4. Number of changes in income/ expenditure structures	.683*	.547*	.181
5. Number of changes in deployment of unit managements	.333	.384**	.295
6. Number of changes in core councils	0.0	.529*	.417*
7. Number of changes in super substructures	.331	.497*	.371**

* Significant at 1 per cent level of confidence.
** Significant at 5 per cent level of confidence.

tion was entirely different at the time of the second set of *kartas* when they were significantly related with all seven variables. Again, at the time of the first set of *kartas*, deaths and retirements were related with the changes in the investment structures, super substructures and core councils, all at the family level. In all but one other case, the coefficients were positive though statistically insignificant.

Adulthood reached is found to be highly correlated with the changes in the investment structures, income or expenditure structures at the family level, and unit level formal management throughout (Table 10.7). They are also significantly related with the number of new units established and the number of major expansions or diversifications effected by the industrial families throughout the historical continuum. In the case of changes in the core councils and super substructures, they are significantly related at the time of the third set of *kartas* and third and second sets of *kartas*, respectively. The coefficients are positive in all other cases but statistically insignificant.

Marriages also show very high and statistically significant

TABLE 10.7 : **Correlation coefficients between adulthood reached and changes in management structures**

Family Management Structure Variables	Coefficients		
	Karta III	*Karta* II	*Karta* I
1. Number of new units	.380**	.701*	.412*
2. Number of major expansions/ diversifications	.418*	.521*	.586*
3. Number of changes in investment structures	.968*	.919*	.908*
4. Number of changes in income/ expenditure structures	.924*	.873*	.586*
5. Number of changes in the deployment of family members in unit managements	.508*	.788*	.508*
6. Number of changes in core councils	.410*	.184*	.287
7. Number of changes in super substructures	.564*	.582*	.288

* Significant at 1 per cent level of confidence.
** Significant at 5 per cent level of confidence.

correlation with six out of seven variables considered in the analysis (Table 10.8). Only in the case of changes in the core councils, are the coefficients not statistically significant at the time of the third and second set of *kartas*.

With a few exceptions, therefore, the second main hypothesis is validated.

RELATED HYPOTHESIS

The total own size of the industrial families and the number of males are likely to be related to the number and size of the management structures at the family level, as well as with the number of economic units

On empirical testing, the total own size of the families and the number of males are found to be highly correlated with the number of investment and income or expenditure structures, sizes of investment and income or expenditure structures, super substructures, and formal managements of the economic units and

TABLE 10.8 : **Correlation coefficients between marriages and changes in management structures**

Family Management Structure Variables	Coefficients		
	Karta III	*Karta* II	*Karta* I
1. Number of new units	.471*	.680*	.410*
2. Number of major expansions/ diversifications	.445*	.617*	.596*
3. Number of changes in investment structures	.624*	.862*	.784*
4. Number of changes in income/ expenditure structures	.556*	.819*	.596*
5. Number of changes in the deployment of family members in unit managements	.548*	.765*	.489*
6. Number of changes in core councils	.318*	.262	.424*
7. Number of changes in super substructures	.532*	.591*	.396**

* Significant at 1 per cent level of confidence.
** Significant at 5 per cent level of confidence.

number of economic units with the industrial families (Tables 10.9, 10.10). Core councils are not included in this exercise since, as explained earlier, these are compact and elitist structures.

THIRD HYPOTHESIS

The nature of the authority structures likely to determine the presence or absence of turbulence in the family

On empirical testing, attention control is *not* found to be related to the general familial turbulence, turbulence in the domestic and economic spheres and total familial turbulence (except at the time of the first set of *kartas* when there is significant correlation in case of general familial turbulence, turbulence in economic sphere and total familial turbulence). No causatory relationship exists between attention control and the four types of turbulence at the time of the second set of *kartas* (Table 10.11). It, therefore, emerges that attention control caused turbulence in the family and economic spheres only in such cases where it

TABLE 10.9 : **Correlation coefficients between family size and management structures**

Management Structure Variables	Coefficients		
	Karta III	*Karta* II	*Karta* I
1. Number of investment structures	.565*	.564*	.441*
2. Number of income/expenditure structures	.561*	.550*	.524*
3. Size of investment structures	.979*	.951*	.978*
4. Size of income/expenditure structures	.978*	.957*	.983*
5. Size of super substructures	.568*	.524*	.635*
6. Size of formal management of the units	.938*	.892*	.898*
7. Number of units	.430*	.465*	.367**

* Significant at 1 per cent level of confidence.
**Significant at 5 per cent level of confidence.

TABLE 10.10 : **Correlation coefficients between number of males and family management structures**

Manegement Structure Variables	Coefficients		
	Karta III	*Karta* II	*Karta* I
1. Number of investment structures	.588*	.580*	.407*
2. Number of income/expenditure structures	.579*	.565*	.511*
3. Size of investment structures	.923*	.655*	.869*
4. Size of income/expenditure structures	.924*	.662*	.878*
5. Size of super substructures	.483*	.372**	.500*
6. Size of formal managements of the units	.980*	.664*	.963*
7. Number of units	.481*	.515*	.333

* Significant at 1 per cent level of confidence.
** Significant at 5 per cent level of confidence.

had become intolerable.

Person control, on the other hand, is significantly correlated

TABLE 10.11 : **Inter-correlations between nature of anthority structure and turbulence in the industrial families**

Set of Variables		Coefficients		
Correlation of variables	With variable	*Karta* III	*Karta* II	*Karta* I
X^1	X^6	.143	0.0	.464*
	X^7	.133	0.0	.264
	X^8	.056	0.0	.393**
	X^9	.136	0.0	.410*
X^2	X^6	.599*	.640*	.464*
	X^7	.489*	.451*	.264
	X^8	.244	.380**	.393*
	X^9	.546*	.562*	.410*
X^3	X^6	.898*	.891*	.881*
	X^7	.477*	.369**	.435*
	X^8	.204*	.172	.490*
	X^9	.633*	.449*	.716*
X^4	X^6	.104	.190	.457*
	X^7	.839*	.898*	.789*
	X^8	.875*	.861*	.547*
	X^9	.835*	.851*	.647*
X^5	X^6	.591*	.723*	.658*
	X^7	.645*	.669*	.540*
	X^8	.452*	.583*	.564*
	X^2	.715*	.773*	.682*

Note : X^1=Attention control ; X^2=Person control ; X^3=Attributed authority in domestic sphere ; X^4=Attributed authority in economic sphere ; X^5=Aggregated authority structure index ; X^6=General familial turbulence ; X^7=Turbulence in domestic sphere ; X^8=Turbulence in economic sphere ; X^9=Total familial turbulence.

* Significant at 1 per cent level of confidence ;

** Significant at 5 per cent level of confidence.

with all types of turbulence except with general familial turbulence at the time of the first set of *kartas* and with turbulence in the economic sphere at the time of the third set of *kartas*. The coefficients are positive even in these two exceptions although not statistically significant. Attributed authority in the domestic sphere is also significantly correlated with all types of turbulence in the family except with turbulence in the economic sphere at the time of the second and third sets of *kartas*. The coefficients are positive in the case of the two exceptions although satistically insignificant.

Attributed authority in the economic sphere too is significantly correlated with all types of turbulence in the family except with general familial turbulence at the time of the second and third sets of *kartas*. Even in these two exceptions the co-efficients are positive although statistically insignificant. The aggregated authority structure index is also significantly correlated throughout with all four types of turbulence.

The third hypothesis is validated in the case of person control, attributed authority in the domestic and economic spheres and aggregated authority structure index, but rejected in the case of attention control at two out of three points of time.

Related hypothesis

1. Turbulence of various kinds are likely to be interrelated

As hypothesized, generally, turbulence of various kinds are significantly correlated (Table 10.12). There are only three exceptions when this interrelationship is positive but statistically insignificant. These cases are, (i) general familial turbulence with turbulence in the domestic and economic spheres at the time of the second set of *kartas* and (ii) general familial turbulence with domestic turbulence at the time of the first set of *kartas*. Only in one of these three exceptions is the coefficient really very low (.068).

With three exceptions, the related hypothesis is validated.

2. Higher turbulence is likely to be accompanied by highly effective strategies employed to combat turbulence

Empirically, highly effective strategies were indeed found to have been employed to combat various types of turbulence (Table 10.12). Out of 12 possible pairings, the relationship is positive but statistifically not signicant only in two cases. These relate to general familial turbulence and turbulence in the economic sphere.

The hypothesis is generally validated.

TABLE 10.12 : **Inter correlations between familial turbulence and effectiveness of strategies employed to combat turbulence**

Variables	Coefficients				
	X^1	X^2	X^3	X^4	X^5
			Karta III		
X^1	—	.520*	.489*	.549*	.617*
X^2		—	.689*	.876*	.680*
X^3			—	.842*	.399*
X^4				—	.709*
X^5					—
			Karta II		
X^1	—	.215	.068	.421*	.521*
X^2		—	.848*	.910*	.580*
X^3			—	.894*	.417*
X^4				—	.583*
X^5					—
			Karta I		
X^1	—	.349**	.332	.672*	.238
X^2		—	.551*	.753*	.653*
X^3			—	.847*	.328
X^4				—	.507*
X^5					—

X^1=General familial turbulence
X^2=Turbulence in domestic sphere
X^3=Turbulence in economic sphere
X^4=Total familial turbulence
X^5=Effectiveness of strategies employed to combat familial turbulence
* Significant at 1 per cent level of confidence.
** Significant at 5 per cent level of confidence.

FOURTH HYPOTHESIS

The business performance of the industrial families is likely to be dependent on the effectiveness of the strategies employed to combat familial turbulence, and to control internal and external business environments

Four effectiveness variables, namely, effectiveness of strategies employed, (i) to combat familial turbulence, (ii) to control the internal business environment, (iii) to control the external business environment and (iv) to control the total business environment were correlated with three measures of business performance, namely, (i) own wealth, (ii) managed wealth and (iii) total wealth (Tables 10.13, 10.14, 10.15). Of these, the last three are significantly correlated with each of the three measures of business performance throughout the historical continuum. Effectiveness of strategies employed to combat familial turbulence is positively correlated throughout, but significantly only once (at the time of the first set of *kartas* in the case of managed wealth).

To find out if some other groups of variables also had any significant relationship with business performance, nine variables concerning the demographic structure and changes in it, and

TABLE 10.13 : **Correlation coefficients between own wealth and effectiveness of strategies combating turbulence, nature of authority structure and demographic variables**

Independent Variables	Coefficients		
	Karta III	*Karta* II	*Karta* I
1. Effectiveness of strategies combating familial turbulence	.163	.163	.262
2. Effectiveness of strategies controlling internal environment	.509*	.536*	.442*
3. Effectiveness of strategies controlling external environment	.427*	.435*	.599*
4. Effectiveness of strategies controlling total business environment	.470*	.480*	.583*
5. Attention control	.002	0.0	—.187
6. Person control	—.216	—.278	—.187
7. Attributed authority in domestic sphere	–.116	–.152	–.268
8. Attributed authority in economic sphere	.211	.092	.075
9. Aggregated authority structure index	–.051	–.154	–.182
10. Family size	.307	.394**	.325
11. Adulthood reached	.367**	.496*	.373*
12. Marriages	.372**	.466*	.383*
13. All demographic changes	.389**	.463*	.407*

* Significant at 1 per cent level of confidence.
**Significant at 5 per cent level of confidence.

TABLE 10.14 : **Correlation coefficients between managed wealth and effectiveness strategies of combating turbulence, nature of authority structure and demographic variables**

Independent Variables	Coefficients		
	Karta III	*Karta* II	*Karta* I
1. Effectiveness of strategies combating familial turbulence	.182	.222	.371**
2. Effectiveness of strategies controlling internal environment	.525*	.550*	.473*
3 Effectiveness of strategies controlling external environment	.433*	.459*	.627*
4. Effectiveness of strategies controlling total business environment	.480*	.500*	.615*
5. Attention control	–.016	0.0	–.163
6. Person control	–.221	–.252	–.163
7. Attributed authority in domestic sphere	–.114	–.096	–.289
8. Attributed authority in economic sphere	.177	.092	.134
9. Aggregated authority structure index	–.069	–.098	–.158
10. Family size	.369**	.384**	.366*
11. Adulthood reached	.427*	.480*	.363**
12. Marriages	.434*	.457*	.399**
13. All demographic changes	.450*	.451*	.443*

* Significant at 1 per cent level of confidence.

** Significant at 5 per cent level of confidence.

the nature of the authority structure were correlated with business performance (Tables 10.13, 10.14, 10.15). Among these, the five variables indicating the various dimensions of the nature of the authority structure, in general, expectedly show low and insignificant inverse correlation coefficients. It is evident that the nature of the authority structure is better reflected in the strategies employed to combat turbulence and to control the internal and external business environment.

On the other hand, family size (except at the time of first and third sets of *kartas* in case of own wealth), adulthoood reached, marriages and all demographic changes are throughout

TABLE 10.15 : **Correlation coefficients between total wealth and effectiveness strategies of combating turbulence, nature of authority structure and demographic variables**

Independent Variables	Coefficients		
	Karta III	*Karta* II	*Karta* I
1. Effectiveness of strategies combating familial turbulence	.178	.200	.321
2. Effectiveness of strategies controlling internal environment	.523*	.547*	.462*
3. Effectiveness of strategies controlling external environment	.432*	.452*	.618*
4. Effectiveness of strategies controlling total business environment	—.479*	.495*	.603*
5. Attention control	—.012	0.0	—.175
6. Person control	—.220	—.264	—.175
7. Attributed authority in domestic sphere	—.175	—.119	—.218
8. Attributed authority in economic sphere	.185	.125	.107
9. Aggregated authority structure index	—.065	—.121	—.170
10. Family size	.355*	.391**	.349**
11. Adulthood reached	.414*	.489*	.369**
12. Marriages	.421*	.466*	.394**
13. All demographic changes	.438*	.459*	.429*

* Significant at 1 per cent level of confidence.
** Significant at 5 per cent level of confidence.

significantly correlated with all three measures of business performance. Thus, the family demographic structure and changes in it indeed show a significant relationship with business performance. However, in so far as deployment of the demographic resources of the family and adjustment of demographic changes in various management structures is concerned, it is being done by the *karta* and the collective authority structures. As the four effectiveness strategies themselves are highly interrelated (Table 10.16), it is clear that there is a possibility of spurious absence of significant correlation in the case of effectiveness of strategies employed to combat familial turbulence. In order to present a

TABLE 10.16 : **Inter-correlations between effectiveness of strategies employed to combat familial turbulence and to control business environment**

Set of Variables		Coefficients		
Correlation of Variable	With Variables	*Karta* III	*Karta* II	*Narta* I
X^{10}	X^{11}	.444*	.312	.661*
	X^{12}	.542*	.374**	.497*
	X^{13}	.516*	.357**	.630*
X^{11}	X^{12}	.904*	.939*	.642*
	X^{13}	.963*	.976*	.886*
X^{12}	X^{13}	.985*	.991*	.924*

Note : X^{10}=Effectiveness of strategies employed to combat familial turbulence
X^{11}=Effectiveness of strategies employed to control internal business environment
X^{12}=Effectiveness of strategies employed to control external business environment
X^{13}=Effectiveness of strategies employed to control total business environment.
* Significant at 1 per cent level of confidence.
** Significant at 5 per cent level of confidence.

correct picture about the relative contribution of all the effectiveness variables, partial and multiple correlation coefficients were used. In attempting such an exercise, however, the fourth effectiveness variable, namely, effectiveness of strategies employed to control total business environment was dropped because it was an aggregate of the second and third type of effectiveness (Table 10.17). Two definitive conclusions emerge from Table 10.17.

1. The three variables, namely, effectiveness of strategies employed to combat familial turbulence, and effectiveness of strategies employed to control internal and external business environments, respectively, taken as one cluster, explain the performance of the industrial families between 75 and 80 per cent.

2. Among the three effectiveness variables, the one which helps to explain the maximum percentage of variance is the effectiveness of strategies employed to combat familial turbulence. Conversely, the effectiveness of strategies employed to control the external environment emerges as a poor explanatory variable.

TABLE 10.17 : **Partial and multiple correlation coefficients of performance of industrial families with effective variables**

Time Frame	Partial Correlation of X^4			Multiple Correlation of X^4 with X^1, X^2, X^3
	X^1	X^2	X^3	
		Own Wealth		
Karra III	.49251*	.40509**	—.0698	.74531*
Karta II	.52352*	.41736*	—.09229	.77146*
Karta I	.36887**	.55816*	—.00383	.76477*
		Managed Wealth		
Karta III	.50341*	.40322**	—.05908	.75936*
Karta II	.51727*	.41371**	—.04103	.78567*
Karta I	.33228**	.55185*	.10843	.80178*
		Total Wealth		
Karta III	.50266*	.40407**	—.06236	.75767*
Karta II	.52242*	.41769*	—.06326	.78252*
Karta I	.35405**	.55920*	—.05089	.78714*

Note : X^1=Effectiveness of strategies employed to combat familial turbulence.
X^2=Effectiveness of strategies employed to control Internal environment.
X^3=Effectiveness of strategies employed to control external environment.
X^4=Business performance (measured separately through Own Wealth, managed wealth and total wealth).
* Significant at 1 per cent level of confidence.
** Significant at 5 per cent level of confidence.

The partial and multiple correlational analysis thus puts the relative importance of the three effectiveness variables in proper perspective.

11

CONCLUSIONS
Entrepreneurial Character and Class Relations

ONE of the stated objectives of this study was to identify on the basis of empirical findings, such characteristics of family managements in India which could be profitably used by other professional organizations. The study has revealed that economic enterprises are affected by the demographic and authority structures of the industrial families. The degree of influence exercised by them is so considerable that the success or failure of economic enterprises is to a large extent determined by them. While family managements have genuine negative aspects, they also have a large array of positive areas of excellence. From a critical screening of the empirical material of this study, the following have been identified as the areas of excellence worthy of transplant and emulation in any work organization irrespective of the locus of ownership and control.

Preservation of Corporate Entities

One of the significant findings of this study is the presence of very high degree of turbulence of various kinds among the industrial families. However, barring a few cases, all others have successfully preserved their corporate entities. This has been accomplished by deftly handling succession and disputes pertaining to the running of economic units and the efficient handling of all other types of turbulences in the family and environment.

In most cases, the succession issue has not been left to be decided as it has arisen, it has been planned in advance. In fact the imaginative feature of the planned succession is that retiring *karta* generally remains on the scene with the one succeeding him during the transitional period. This enables the successor to learn the ropes and respond to various components of the environment under the watchful eyes of the retiring *karta*. Thus, the family as well as the business environment are adequately prepared to adjust to the new *karta*, as is the latter to adapt himself to the demands of his new role. In many of the public sector enterprises in India, top executive positions remain vacant for long durations, and potential successors to the top positions are seldom groomed within the same organization. The practice of Indian industrial families could profitably be emulated in them.

Various members of industrial families informally and formally look after generally allocated economic units. At the family level, a superstructure overviews the operations of the economic units on the lines determined by it. Therefore, it is possible to correct or control the family members if they begin to run the allotted units on the basis of their personal whims and fancies. This strategy does not necessarily enable the economic units always to be run on strictly professional or rational grounds. Nevertheless, it ensures that the interests of the family do not suffer in the bargain. In contrast, the overviewing mechanisms in the public sector enterprises are anonymous entities where personalized care of areas and units is generally absent. This lack of personalized attention to economic units quite often adds to the endemic problem of delays in strategic decision-making, especially quick directions to be given in abnormal or difficult situations. There is a strong case for making specific persons responsible for sharpshooting various types of problems being faced by the public sector enterprises.

The preservation of the corporate entities of the industrial families is not only an emotional issue with their family members ; it is equally an economic issue since remaining together vests in them distinct economic advantage such as tax concessions, xtra-resourcefulness in exploiting business opportunities, an so on. In the public sector enterprises, functional managers operate under the assumption that no threat exist to corporate entities of the units being run by them. Quite a few among them also feel that they have no major personal or emotional stake in the working results. Organizational costs consequently increase. While the units themselves continue to exist, their economic viability suffers. There is, therefore, need of the personal involvement of top managers in ensuring the preservation of the economic health of their units.

Specialized management of turbulence

Although only about 30 per cent of family members posseseed professional or technical qualifications to run economic units, the families themselves more than made up for it by managing turbulence in the family and in the economic sphere in a very specialized manner. The management structures at the family and economic unit levels generally are also turbulence management structures. In fact, some among them exist and work only for this purpose.

One of the most effective attributes of the organizational structures used by the industrial families is the compactness of the management superstructure. Although the size of total wealth has been expanding and the number of economic enterprises increasing over the years, the size of management superstructures, specifically typified by the core councils and substructures, has not been increasing in the same proportion. These are informal structures and have been working without any legal or formal existence after the abolition of the managing agency system. Nevertheless, they are the mechanisms by which strategic issues are instantly sized up, quick decisions taken, resources allocated or relocated, and direction of further action to be taken inside the work organization and external environment very precisely indicated. There is no system-based culture of decisional inertia—to take a decision or not to take a decision—as in government departments or public sector enterprises. This dynamism, sense of direction and unity of purpose has to be given to any complex work organization if it is to operate on the principles of rationality, natural justice, fair play and effectiveness.

Although the total number of management hierarchies reported was six, in reality the structural distance between them has been greatly reduced by effective integration and overlap of membership. First, for example, the membership of the core councils and the substructures, overlaps ; the membership of the latter overlaps among its many structures. Second, the formal management at unit level also have a mix of the core council, substructure members, and unattached family members. Third and more importantly, this direct involvement link is extended even to understructure levels, providing an additional option to managements to combat turbulence. Since most members who belong to these management hierarchies are family members, strategic issues can even be discussed at breakfast, lunch and dinner tables without any formal, procedural wrangling. Reduction of distance between hierarchies avoids quite a bit of organizational anxiety, unnecessary bureaucratization of decision-making, and reduces the time otherwise spent on meeting procedural formalities.

The substructures, which are specific turbulence management structures, are useful in both types of choices made by the core councils. In the case of sound choices, these structures help in maximizing gains. In the case of sub-optimal choices, they minimize losses. To the extent that these structures manage turbulence speedily, they reduce any potential damage that can otherwise be caused to the families and/or economic organizations.

The industrial families singly and as a class possess a large corpus of effective strategies for combating internal and external turbulence. For example, it is part of modern Indian history that they had all done fairly well under the British regime which had turned the Indian economy into a hapless dependency of the colonial power. In the post-independence period, they forged very close links with every single government that was in office at the Centre. Each of these governments pursued policies that were, if one were to go by the proceedings of organizations such as FICCI and ASSOCHAM, almost 'totally detrimental to their interests.' Despite constantly crying a wolf—perhaps that too as a strategy—they have all recorded impressive growth. It is true, of course, that quite a large number of their effective strategies, used to obtain the results they have succeeded in achieving, are simply manipulative and cannot be recommended for emulation by professionally managed organizations. However, it is equally true that they also have clean strategies that can be accepted by others. These are, (i) quick articulation of the problem and effective mobilization of 'distinctive competence', (ii) detailed identification of support and 'enemy' strengths and weaknesses and (iii) constant readiness to accept something unavoidable and making a quick adjustment with it.

Satisfaction of the needs of family members

Through their development and deployment in economic enterprises, the industrial families satisfied the following five important ego needs of their family members :

1. Their identity needs are fulfilled through their affiliation with specific economic units.

2. Their status needs are satisfied primarily by their placement in various positions on the management structures of family and unit levels, that is, core councils, substructures, and unit level managements. These placements decide the seniority and importance not only inside the family but also in the outer business environment.

3. The family members are assigned specific roles. The members of the core councils and the substructures generally figure prominently in strategic decision-making. Others participate in the decision-making process through consultation about 'their' economic units. They are also given a relatively free hand in implementing decided strategies and taking day-to-day administrative and operative decisions.

4. Family members gain prestige through the performance of their assigned tasks, and its extent is generally associated with the importance of the given task.

5. The existent reward system enhances the status of the rewarded family members and quite often also fulfils some of their cherished desires. Because the ego needs of the family members are generally satisfied, it helps family managements to combat turbulence in their families. It also enables them to reduce the possibility of family members using their energy to work against the interests of the families and each other.

Effective deployment of familial resources

In order to obtain the maximum possible gains, the industrial families develop and deploy human and financial resources in very effective ways. Development of human resources takes the following ways : Male members of the families are watched *vis-a-vis* their special talents, aptitude and aspirations by members of the core councils and the *kartas*. As they grow up, they are groomed in stages for specific roles in the economic enterprises controlled by the families. These stages are, apprenticeship, co-authorship, partial autonomy, and fuller autonomy. This type of entrepreneurial development is natural and quite fast. Learning is devoid of excessive theoretical padding, and any element of error by the understudies is overviewed by the role models. The potential for commandeered resources is increased through the process of corporate marriages. This helps in consolidating their interests, increases the number of their allies and reduces militant competition.

The development and deployment of economic resources involves the working out of an elaborate mix of various segments of the economic resources. The industrial families employed 10 sequential strategies in deploying their economic resources in terms of (i) immovable assets, (ii) cash and semi-liquid assets, (iii) investment in business and trade and (iv) investment in the manufacturing industry. These strategies reflect shifts in emphasis on different segments of the resource structure activated at the time of the various sets of *kartas*. In terms of total returns from their economic units and the number of products covered, the performance of the industrial families has been fairly impressive from their points of view.

NEGATIVE ASPECTS OF FAMILY MANAGEMENTS

Although it was not one of the stated objectives of this study to identify the negative aspects of family managements, it is considered useful to do so for two reasons : first, to give a more balanced picture of empirical realities, the negative aspects certainly need to be fully taken note of ; and second, given political will, there is a strong case effectively to curb them. The negative aspects include the following :

Investment techniques

Indian industrial families have adopted a very complicated system of investment which makes it extremely difficult to identify the real identity of the investors. This serves two purposes :

1. Owing to non-detection of their identity, industrial units can avoid regulatory rigours under the MRTP Act ;

2. Black money can be invested under false names and control kept in a few hands.[1] Contributory factors for the wide-

spread use of these devices are the industrial regulation and taxation laws that give unnecessary advantages to what are called Hindu Undivided Families and the trusts on the one hand, and partnerships, etc. on the other. At a time when it is more than clear that these laws are being used for tax avoidance and manipulation, it is difficult to understand why they are not being changed.

The absence of family-wise registration of companies gives rise to the suspicion of perhaps the not so concealed connivance of the bureaucracy and even the politicians in not implementing the provisions of the MRTP Act. Under this Act a key requirement for processing of the applications from monopoly houses is the identification of the links of the applicant. In fact the very reference of applications to the MRTP Commission is a patently partisan political decision which is applied in some cases and not in certain others. Khurana (1979) indicates how the bigger industrial families have thrived in spite of the existence of the MRTP regulations. The commission itself has indicated, on numerous occasions since its existence, why it has not been able effectively to curb the existence of monopolies on the one hand, and unfair trade practices on the other.

Structure of employor-employee relations

The structural inequality of employer-employee relations is such that it is not possible for the workers to ensure fair wages for themselves through collective bargaining. This is partly so because of the fact that the state as a mediator generally sides with the employers through system-inertia, non-intervention, loopholes in legislations, and non-implementation of the existing statutory provisions. There is a strong case for taking wage fixation out of the hands of private industry and determine it through legislation across the nation in a wider framework of national income and wages policy. Similarly, since the workers contribute very significantly to the industrial sector, their participation in the management of enterprises especially strategic decision-making involving design of work environment, career development etc. needs to be statutorily defined. Functional managements, paid for by shareholders and consumers but working for the promotion of the interests of only the industrial families, need to be made equally responsive to the interests of the former. Within the existing framework, their role is defined by the industrial families. Once their role and responsibilities are statutorily defined, given protection, allowed adequate role-space, and functional freedom to operate, it would be possible for them to salvage their identity. Similarly, in retrun for fair wages, productivity norms of the workers could well be linked with incentive payments.

Financial control

One single functional area of management of enterprises in which the industrial families in India have left very little elbow room for the managers to do is financial control. In the process, company funds are quite often diverted for other purposes. Diversion of funds sometimes leaves inadequate working capital and this ultimately contributes to turbulence. Quite often, the units get closed or are forced to stop payment of wages and dues to other agencies. To tide over the crisis they take huge loans from industrial financial institutions and commercial banks. Under pressure, the government asked financial institutions to put convertibility clauses in the loan agreements exceeding certain limits which entitle them to convert the loans into shares and to nominate their nominees on the boards of directors. The industrial families have been clamouring for its removal ever since its imposition. Some among them have even preferred to close a unit rather than taking the trouble of inviting nominee directors on their boards.

The government has been well aware of the techniques of financial manipulation and the sickness which such practices quite often cause to industry. Whenever the issue has cropped up in Parliament, the government has always "threatened" to introduce legislation which would empower it to intervene without taking over the management of the concerned undertakings. For example, in October 1976 an inter-ministerial committee was constituted to prepare a comprehensive legislation to :

(a) deal with sickness in industries in the early stages by creating an information system to monitor industrial undertakings at various stages of incipient sickness and to take remedial measures ;

(b) create special cells in financial institutions to pick warning signals through an analysis of what are called 'critical ratios';

(c) create a special cell in the Reserve Bank of India to play a co-ordinating role and to serve as a clearing house for information emanating from banks before it is communicated to the government for remedial purposes; and

(d) enact a statutory provision under the Industries (Development and Regulation) Act 1951 to authorize governmental intervention in the management of sick units. No effective follow up action was taken after the committee presented its report.

Another related aspect was encouragement given to mergers of sick units with sound ones. In 1976, a working group constituted specifically on this issue recommended that:

(a) mergers between the sick and sound units are likely to bestow upon the former the benefits of "better management" and help them to restore confidence in their future viability;

(b) to encourage such mergers it is essential to provide some tax concessions and waive certain provisions of MRTP and Companies Acts; and

(c) a package of measures to encourage potentially viable mergers is deemed to be in 'public interest'.

These recommendations were incorporated in a statute in the Finance Bill of 1978-79. Of course, the motive behind such recommendations is very clear: the larger industrial families that have created several such sick units themselves and are not interested in nursing them back to sound economic health, are to benefit even through the take-over of the units. There is also the possibility of the same industrial families absorbing the profits of its viable with its non-viable or sick units with attendent tax benefits to boot.[2]

Mr. George Fernandes, during his tenure as the Minister of Industries had also floated the idea of a fourth industrial sector (apart from the private, public and joint), namely, the workers, sector which could have enabled the workers to run sick units. However, the Ministry of Industry did not take any steps on the establishment of a management cadre among the financial institutions to manage sick companies. One more inter-institutional group headed by Mr. Bhuchar of the Reserve Bank of India, had deliberated on the issue and presented its report on 22

January 1979.

By 1985 the attitude of the Indian government on industrial sickness had undergone a significant change. From its previous policy of taking over management of sick units of private sector primarily to avoid unemployment of the large number of workers, nationalization, etc. it had veered round to quietly allow a large number of economic units of some industries (i.e. sugar, textile, jute,) to die. Wherever it had succeeded to nurse the sick units baek to their health, it was poised to hand them back to their original owners. There were also indications of pushing through a policy of phasing out a significant segment of workers from economic units employing a new technology which rendered quite a few existing hands of workers redundant.

Role of industrial families and government in industrial sphere

In the absence of a clear-cut, well-defined government policy about production capacities, technology-use, product-mix and pricing, public financing and intervention mechanisms, the role of industrial families and the government in the industrial sphere has been reduced to one of using the other as the occasion demands. It is not that the industrial families have the same level of freedom in industrial governance for all their industrial units ; it greatly varies from one unit to another depending upon tho nature of the product, participation of financial institutions in its investment structure, its linkages with the other product groups in the economy and its market. To the extent that the role-space of the government increases, it reduces the role of the industrial families, Thus, at one end of the continuum are companies such as the Tata Iron and Steel Company (TISCO) for which the government decides the product mix, pricing, synchronization with other public sector steel plants, issue of shares and the emoluments of wholetime directors. Most strategic issues are out of the hands of the Tata family in the management of TISCO and to that extent its role is reduced and made less consequential. At the other end of the continuum are such companies where almost every strategic issue is in the hands of the Tata family and where the role of the government is but minimal. In the interaction of the industrial families and politicians, however, this crucial question somehow recedes to the background and the two sides, instead of respecting each other's territory, have attempted to depict the issues in a rather partisan manner.

Thus, theoretically both the government and the industrial families can influence each other a great deal. In the actual exercise of influence, however, it is evident that generally the industrial families use the government—at least the key segments of it—and not vice-versa. Overall, this easy vulnerability of the government has culminated in highly uneven industrial development in the country.

The role of industrial families in India has been a live issue among the political parties. During the Janata regime (1977-1980) its Executive had asked the government on 26 March 1978 to take immediate steps to.

(a) decentralize the ownership of means of production ;

(b) impose restrictions on the chairmanships, managing directorships of a large number of corporate companies by a few individuals ; and

(c) bring corporate units under the control of public financial institutions in those instances where they have provided a large percentage of overall capital, and curb the malpractice and exploitation.

Similarly, on 30 March 1978 the then industries Minister, George Fernandes, had indicated the 'intention' of the government to delink the ownership of companies from big business families through a proposal which would have every member of an industrial family separately own or control one or two companies.

By October 1978 the government had received the report of the Sachar Committee which, among other things, had recommended curbing of the growth of large industrial houses. The Ministry of Industrial Development, in a note circulated to the Janata Party Members of Paliament, had given an indication on 5 October 1978 that government would accept most of its recommendations by :

(a) permitting in future inter-corporate investment and loans "only in the interest of the growth of the economy" and "only in certain specified categories of companies" ;

(b) redefining investment companies to mean "companies limited by shares and carrying on business of only underwriting in or dealing in shares, debentures or other securities" ;

(c) restricting the expansion of dominant undertakings ; and

(d) enlarging the role of the public sector.

This move was followed by a decision to bar the entry of big industrial families in a few listed industries in December 1978 and a resolve to take over key industries such as steel, aluminium and automobiles as a part of a comprehensive plan to check the growth of economic power.

Meeting in the background of the proposal to nationalize the key industries, the Janata Party National Executive had on 18 January 1979 neatly skirted the nationalization issue while reaffirming its "commitment to decentralize economic power and promoting healthy industrial development by reducing the control and holding of large industrial houses." It had authorised the party president to nominate a committee to study the question and make recommendations. This committee met several times and, encountering pulls from several quarters, considered a number of proposals—including some from the late Professor Raj Krishna. As usual nothing came out of it.

The debate has continued in the same vein in the post 1980 phase and may do so even in future. It is certain, however, that the industrial families would not be altogether eliminated from the industrial scene. Since the public sector has shown its own strengths and weaknesses, a crucial area of experimentation is the joint sector concept. Erdman (1973) and Murthy (1979) have provided differing assessments of different variants of the joint sector. Erdman (1973, pp. 128-29) after examining the Gujarat State Fertilizer Company (GSFC), a joint sector enterprise with private management control but public funds and accountability, praised the experiment. He argued that joint sector companies must be allowed to function as private sector companies and the key to autonomy lay in the voluntary abstinence of the government from actively intervening in these enterprises. Murthy (1978), on the other hand, pointed out that such experiments resulted in :

(a) control of large projects by the private sector with relatively little investment of their own ;

(b) entry into industries ordinarily out of bounds for the private sector ; and

(c) obtaining concessions not otherwise available. Since the state industrial development corporations possessed tenuous organizational resources to exercise any effective degree of management control and since more of such projects were selected on the rationale of "not being left behind," he argued, this had quite often led to industrial sickness via the now familiar route of spiralling project costs, low capacity utilization, and inability to market the product.

In the days to come all these issues will have to be sorted out by the government if it is genuinely interested in promoting balanced industrial development.

INDIAN INDUSTRIAL FAMILIES AND INTER AND INTRA-CLASS RELATIONS

Discussion of issues such as class formation, class composition and inter-and intra-class relations in India has suffered from two inadequacies : one, conceptual clarity about the terms used and their application in explaining social inequality, exploitation and conflict has generally been lacking, and two, discussion on the question has quite often been reduced to less consequential aspects as, for instance, the never-ending debate on the relationship of caste with class (Bailey, 1957, Beteille, 1965 ; D'Souza, 1967, 1969, 1978) or the ethnic background of the capitalist class (Gadgil, 1951, 1959 ; Lamb, 1955, 1956 ; Mahnot, 1962 ; Pavlov, 1964 ; Timberg, 1969, 1978). The main issues of inter-and intra-class relations and conflict, consequently, get inadequate coverage in these explanations.

It is not proposed to make an ambitious corrective exercise : On the contrary, the attempt here is more modest and confined to a discussion of the formation of the capitalist class, in particular the role played by the family, its composition and inter-and intra-class relations in the background of data generated by this study. Before commencing the exercise, however, it is essential to specify the conceptual and theoretical position taken on the question.

Class is viewed in terms of relationship to the means of production. The basic unit of inequality in class analysis is not considered to be an individual, as seen by some (Beteille, 1965 ; D'Souza, 1967, 1969, 1978) but a group (Berreman, 1978, p. 9). In pure compositional terms, the social background of those constituting a class is not viewed to be necessarily and always homogenous. Similarly, members of the same class are structurally not expected always to enjoy the same status. While it is recognized that the initial entry into a class is more or less an achieved one, continuation among certain classes (i.e. capitalist) is via ascribed status, that is, birth. Straight entry into some classes, while theoretically open, is not all that open, there being several intermediate stages involved in reaching it. As a process, class formation is viewed as both spontaneous and organized. In the latter case the time element involved in the class consolidation is considerably reduced. Further, it is felt that conflict occupies the central position in the analysis of class relations.

Interpreting the empirical evidence of the present study in the above theoretical framework, the following conclusions emerge :

1. Apart from the reasons advanced by Gadgil, Lamb, Pavlov and Timberg, the family, as a domestic organization, has played a very crucial role in the formation and consolidation of the capitalist class in India. Capital accumulation process, through means fair and foul, and consequently industrial development in India on the capitalist path would have received a serious setback if the amassed wealth was quickly squandered by the industrial families in leisure time pursuits. As seen earlier, the Indian family as an institution has been an object of unsubstantiated criticism about its alleged antagonism to saving and hindrance to entrepreneurship. Evidence from this study proves how misplaced such criticism is. Industrial families have helped in the acceleration of the process of capital accumulation through their sheer continuation as single corporate entities by :

(a) avoiding fission in spite of the presence of tremendous amount of segmentation and turbulence ;

(b) skillful deployment of human and financial resources of the family ;

(c) non-distribution of familial income to the individual members and centralized control over spending ; and

(d) avoidance of taxes (through various expedients).

Ultimately, it has resulted in the amassed capital being ploughed back into trade, commerce and industry which has in turn have significantly contributed to the process of consolidation of the capitalist class.

2. Pavlov (1964) and Timberg (1969, 1978) have discussed the different routes to capitalism taken by different Indian ethnic

groups (i.e. Marwaris, Parsees, Gujaratis, etc.) among the Indian capitalist class. Carrying this type of analysis one step further, the present study has discovered evidence of some internal ordering of these ethnic groups within the capitalist class. The hierarchies in the internal ordering of these groups are linked with the total wealth acquired as a result of their control over economic activities including industrial units on the one hand, and complexities of the families on the other.

3. Intra-class relations among the Indian industrial families provide evidence of support and conflict within the same class. Indian industrial families co-operate with each other in :

(a) lobbying with the government through their national-level associations such as FICCI, ASSOCHAM, and the like on the general economic and fiscal policies of the government affecting them as an aggregated class ;

(b) consolidating their class by acting as sleeping directors on joint-stock companies controlled by others ;

(c) effecting production cuts, price increases, restricted releases, etc. in a concerted manner to beat consumer interests ;

(d) co-operating on the whole gamut of issues, that determine the exchange relationship with the workers, as a conscious device to keep the bargaining position of the latter at a low ebb ; and

(e) forging a link with multinationals in an even larger effort directed at exploiting internal and external markets.

They cement their close business relationship by entering into corporate marriages and coming together in cornering political control of the local civic administrations directly, and the state apparatus indirectly.

Intra-class differentiation generally breaks along the ethnic group lines. At highest levels this differentiation found its expression in :

(a) the formation of opposing national associations (FICCI and ASSOCHAM) ;

(b) use of the different political lines (Birla and Tata) ;

(c) in the use of the state apparatus for capacity pre-emption, monopoly and dominance in the market ;

(d) management styles ;

(e) familial identities ; and

(f) legitimation of wealth through the promotion of charitable educational, medical, welfare, cultural and religious institutions.

Structurally, it is possible to trace these conflicting relationships within the capitalist class to its internal ordering (monopoly and non-monopoly capital) and usefulness of one faction to another. The functioning of non-monopoly capital is useful to the monopoly capital in as much as it helps the latter to choose moments of expansion, minimize risks, recoup technological innovations at lower cost, subjecting labour power, price fixation and as a secondary line of production. It is because of these uses that monopoly capital does not brusquely eliminate non-monopoly capital. They coexist and co-operate because it is useful for them to do so. And yet they also clash, for there are inherent contradictions in their relationships.

4. In the overall hierarchy of classes in the Indian economy, the placement of various fractions of the different classes *viz-a-vis* the capitalist class is found more complex than it appears on the surface. Some of the members of the working class (i.e. the trade union representatives) and the new petty bourgeoisie[3] (i.e. the wage-earning groupings) display characteristics favourable to the capitalist class and apparently work against the interests of their classes. Despite their pro-capitalist orientation and behaviour, it would be incorrect to treat them as part of the capitalist class. It needs to be noted by the working class that the strategy of its struggle is in wrong hands and this spurious element needs to be eliminated from the working class organizations to give them a decisive edge in the class struggle.

The managers of the economic enterprises of the capitalist class are performing the functions of the capital and occupy places of capital. Following Poulantzas (1974, pp. 180-81), structurally they can be placed as forming part of the capitalist class. They do not constitute a distinct fraction of the capitalist class since they do not have a relationship or a place of their own and also because there is no separation between the place of capital and its powers.

5. The state apparatus has been primarily responsible for the production and extended reproduction of different classes in Indian society by (i) maintaining the unity and cohesion of the society, (ii) concentrating and sanctioning class domination and

(iii) reproducing ideological and political relations.

INDIAN INDUSTRIAL FAMILY AND COMPLEX ORGANIZATIONS

Reviews on the state of organizational theory (Argyris, 1973; Cashman, et al, 1976; Goldman, 1978; James and Jones, 1976) indicate the inadequacies of the existing explanatory models. Without going into the details of these inadequacies, the discussion here is confined to certain myths that continue to persist about the so-called traditional organizations such as the family. It would be useful to examine and weigh their validity against the backdrop of the evidence earlier presented.

1. One of the prevailing myths is that the work organizations are complex. By implication, the argument could be stretched to mean that traditional organizations such as the family are not. This study has provided ample evidence of the enormous complexity of the Indian industrial families. This complexity is structural and has tremendous implications for other complex (especially industrial) organizations. Structural complexity of the industrial families reflects itself in their total size that sometimes exceeds 80 members drawn from a large number of lineal and collateral segments and sometimes scattered in many locations. Further, the working of the biological processes of production, maintenance and replacement keeps on renewing the membership stock, generating in the process a large number of critical events that affect the working of the family on the one hand, and complex industrial organizations and other components of the external business environment on the other. Even though the Indian industrial families have supposedly reached, at the time of third set of *kartas*, the most complex stage in compositional terms—they were all lineal, collateral joint—they have hardly exhausted the limits of complexity even on this dimension. For it is possible to examine the degrees of their compositional complexity in terms of a number of lineal and collateral segments, independently or even in juxtaposition with their geographical locations. The list of issues involved in the demographic complexity could be increased by adding variables such as generational strength, depth, sex, and so on.

The complexity of the authority structure of the industrial

families is visible as they deal with decisions pertaining to general issues, and allocation and relocation of familial resources—human and financial—in the domestic and economic spheres. The present study indicates that the basic structure of the traditions and values developed over the years by different sets of *kartas* has survived, although it has also undergone certain changes in the process of adjustment with changing requirements. The authority structures, as also the smaller subsystems of the families, have changed somewhat in terms of their styles of operation although their structure has remained almost totally unaltered.

2. The second myth about the complex organizations is that the managers in them create the organizational climate, make and unmake their roles, and that changes in them can be introduced with the help of behaviour intervention techniques that generally modify the attitude and behaviour of the managers, employees and workers. An implicit assumption in this line of argument is that the formal structures have the complete freedom to delineate managerial roles and create an internal climate in the formal organizations. Recent researches in the West, however, indicate that informal understructures, operating below the formal structures play a key role in the managerial role-making process, and that these mechanisms even now very much need to be discovered and mapped before being made effective use of (Dansereau, 1972 ; Graen, 1975 ; Graen, et al., 1977 ; Cashman, et al., 1976).

Evidence from the present study shows the importance of influence, which the family as a corporate organization exercises over the formal organizational structures in industry, on managers and on various other components of the internal and external business environment. In fact, in its totality of operation in the economic field, the industrial families are certainly as complex and extensive as the so-called complex work organizations. Specifically, this study demonstrates that the authority structure of the industrial families, reflecting itself at the family level in the form of informal superstructures called the core councils and substructures and in the formal managements and informal understructures at the unit levels, takes all the strategic decisions. In the process, the roles of the managers are shaped by them. In the performance of professional tasks, the matrix of inter-personal relation gains undue prominence for the managers. To the extent that the managers are handicapped in taking decisions on objective grounds and because their role-space is defined by the families, the scope of the operation of the effect of the managerial behaviour modification techniques/or measures is reduced.

3. A third set of myths in vogue is that :

(a) The goals of the complex work organizations and institutions such as the family are somehow different ;

(b) The demands put by the two cannot be possibly reconciled ; and

(c) In order to make the work organizations viable and effective it is essential to neutralize the impact of socio-cultural variables on the managers and the work force (Kahn 1964 ; Katz and Kahn 1966 ; Cohen, 1973).

In a very powerful exposition, Wallace (1974, pp. 192-94, 205-15) has argued that :

(a) It is the *social order of the institution itself* (in the present case, family) that is of overriding importance in its institutional effectiveness ;

(b) One single explanatory variable for Japanese industrial miracle is the fraternity of enterprises ;

(c) The Japanese fraternal enterprises fulfil the social needs of the employees ;

(d) In a fraternal enterprise both the employers and employees can mutually benefit by ceasing to be artificial mutual adversaries and by becoming partners ;

(e) The adversary relationship between capital and labour is imposed gratuitously, and partnership is the basis for teamwork and accomplishment of any social change ;

(f) From the fusion of the identities of the employers and employees, only the trade unions would be the losers and the management, workers and the stock-holders would benefit ;

(g) Many of the problems that confront us today are caused by the bureaucratic systems. Bureaucracy's—civil or industrial—trouble arose from a coercive social order based on a false view of human motivation in the world. Bureaucracy gets the least out of the people because it *inherits* co-operation and teamwork. Thus, it is forced to become coercive to get results. Coercive bureaucracy systematically inhibits teamwork, alienates employees and inflates costs. It is wrong to assume that no useful alternatives to bureaucracy exist for large organizations. They do exist and in fact work much better cross-culturally.

The effectiveness of the fraternal system is evident not only in Japan. It is true of any country wherever it exists, including United States (Long, 1978) and Africa (Harris, 1971 ; Quarcoo, 1978 ; Wachuku, 1978). From the fractional evidence made avilable by this study on this issue, it is clear that wherever the Indian industrial families have bestowed on the employees the status of a partner, the internal environment of the work organizations has been remarkbly free from stress and strains, productivity has been high, and the need for an industrial bureaucracy has been greatly reduced. Thus, an alternative model for emancipating the work force and increasing the effectiveness of the work organizations is indeed available within the framework of traditional Indian institutions. Further, it is also evident that the grafting of western solutions is neither the best choice nor the only way out.

4. Another myth in circulation is that in designing and operating the work organizations only people, events, functions, technology and environment matter (Allport, 1962 ; Katz and Kahn, 1966 ; Hage and Aiken, 1969 ; Perrow, 1967, 1970). In the present study it has been convincingly demonstrated from detailed analyses of various management structures—core councils and substructures, investment, income or expenditure structures at the family level and the formal managements and the informal understructures at the unit levels—and their interlinkages with input and output structures connected with the industrial units, that ideology plays a very crucial role in the designing and operation of organizations. In fact, since it is also shown that these structures largely decide what the industrial organizations (which are structured and operated on the basis of stated functions and technologies used) have to do and how, it is reasonable to argue that ideology is indeed the overriding consideration. That this interpretation of empirical reality is not confined only to India is proved by the emerging evidence of the presence of familial control of large corporations in the U.K. (Nyeman and Silberston, 1978) and the U.S.A. (Fitch and Oppenheimer, 1970 ; Zietlin, 1974) and the existence there of similar 'control' mechanisms. Thus, if it is indeed a fact that large corporations or organizations are in turn controlled by other organizations which

are essentially products of ideology, there are valid reasons to believe that ideology provides the rationale for the designing and operation of work—especially complex—organizations.

TOWARD A NEW SOCIOLOGY OF ORGANIZATIONS

From the preceding section it is evident that organizational sociology has not adequately appreciated the role of traditional institutions such as the family and the extension of their essentials into or their fusion with the work organizations. The malady that afflicts the sociology of organizations is, however, far more serious than the scope of the comments offered in the last section. Goldman (1978) has indicated how the four schools in organizational sociology (i.e., the institutionalists, comparative, humanists and decision-theorists) have narrowed down their field of enquiry and consequently presented an incomplete and distorted picture of organizational reality. He has identified four major flaws evident in most organizational theories and researches, namely, (i) ahistoricism, (ii) insularity, (iii) elitism and (iv) absence of class analysis. He, therefore, pleads for a new sociology of organizations. This tradition would not ignore the existing body of work in organizational sociology. However, it would certainly re-interpret and reassess its worth. It would address itself, among other things, to the study of the close connection between organizational consultants and the corporate or government executives so as to get to know their intentions, concerns and styles and the ways in which the owners or managers use organizational research (as the managerial socialization process). This tradition would have to rely on such research methods as observation and investigative reporting to tease out data revealing organizational dynamics. It would lead to greater involvement of the researchers in the data collection as against the present practice of treating data collection work as something below the dignity of senior men. Its aim would be to provide the same type of knowledge and information to other segments of the work organization which the present four schools of organizational sociology provide to only one segment, that is, the management.

A very promising stream of research following this new tradition of organizational sociology has emerged in the west (Goldman, 1973 ; Goldman and Van Houten, 1977, 1978 ;

Heydebrand, 1977 ; Van Houten and Lynn, 1974 ; Zimbalish, 1975). The present study is a very modest contribution in the same area from India.

AREAS FOR FUTURE RESEARCH

The preceding section has, in a way, already identified many areas of research in the field of organizational sociology. However, relating this exercise to the issues thrown up by the present study, the following areas would be promising subjects for future research :

1. A comparative analysis of family managements with several segments representing different degrees of fraternalization and partnerhip extended in the work organizations. Of particular interest would be the analysis of the structure of exchange relationships, organizational climate, presence or absence of organizational strife and stress, organizational effectiveness, job satisfaction and participation in the decision-making process by different segments, and the role-space and utility of managers.

2. A detailed historical analysis of the different stages of entrepreneurship pertaining to different product groups and different individuals in family managements. This would yield the insights used by various entrepreneurs to control the internal and external environment in the light of their previous experiences.

3. A comparative analysis of the performance of the industrial families based on different methodologies, that is, the critical ratios, the Karen-Egan scale, Khandwala et. al. scale, and the wealth measures (used in the present study) to find out their strengths and weaknesses in predicting business performance.

4. A comparative analysis of the role-making process for the managers in the family-based, fraternal, public sector, joint sector and multinational work organizations to indicate the system boundaries of the behavioural intervention process.

5. An historical analysis of government policies dealing with the structure of the exchange relationship between the employers and employees, mediating (support, confrontation and intervention) role of the government, industrial regulation especially crisis-management, sickness, nationalization, etc. to juxtapose the public pronouncements of stated state goals and the practised

reality.

Each of the above studies would involve facing considerable amount of difficulties in the collection of empirical data, as happened in the case of the present study. It would require all the expertise of researchers to ferret the facts out to really make a meaningful and comprehensive analysis.

NOTES

1. The former Minister of Industries, George Fernandes, for example, pointed out on 8 March 1979 the existence of 2000 investment agencies maintained by one single family in India. As he emphasized, this was all "legal".
2. This provision of merging sick units with sound ones was apparently invoked by Bennet Coleman Company of the Sahu Jains in late 1978, thus raising a fierce controversy.
3. For a competent delineation of classes and their inter-relationships, see for example, Poulantzas (1974).

BIBLIOGRAPHY

Alford, B.W.E. : 1977 : 'Entrepreneurship, Business Performance and Industrial Development', *Business History*, 19 : 116-33.

Allport, F.H. : 1962 : 'Structural Conception of Behaviour : Individual and Collective, I : Structural Theory and the Master Problems of Socio-Psychology', *Journal of Abnormal and Social Psychology*, 64, : 3-30.

Ansoff, H.I. : 1968 : *Corporate Strategy*, London.

Anstey, Vera 1973 : *The Economic Development of India*, 1952 London.

Argyris, C. : *The Applicability of Organizational Sociology*, Cambridge.

Arnoff, J. and W.D. Crano : 1975 : 'A Re-examination of Cross Cultural Principles of Task Segregation and Role Differentiation in the Family', *American Sociological Review*, 40 : 12-20.

Arora, Satish K. 1968 : 'Pre-empted Future ? Notes on Theories of Political Development', *Behavioural Sciences and Community Development*, 2, 85-120.

Bagchi, A.K. : 1972 : *Private Investment in India* : 1900-1939, Madras.

Bailey, F.G. : 1957 : *Caste and Economic Frontier*, Manchester.

Banfield, E.C. : 1958 : *The Moral Basis of a Backward Society*, New York.

Becker, G.S. 1964 : *Human Capital*, New York.

———1965 A Theory of Allocation of Time, *Economic Journal*, 493-517.

Becker, H. : 1957 : 'Current Sacred—Secular Theory and Its Development', in H. Becker and A. Boskoff (eds.), *Modern Sociological Theory in Continuity and Change*, New York.

Benedict, B. : 1968 : Family Firms and Economic Development, *South-West Journal of Anthropology*, 24.

Berreman, G.D. : 1978 : 'Caste, Class and State in Peasant India, paper presented to the session on Untouchability in the Asian Countries' x *International Congress of Anthropological and Ethnological Sciences*, 15 December, New Delhi.

———1978(a) : Caste and Class, paper presented to the session on Anthropology of Peasantry, post-plenary session of *X International Congress of Anthropological and Ethnological Sciences*, December 21, Lucknow.

Beteille, A. : 1965 : *Caste, Class and Power* : Changing Patterns of Stratification in a Tanjore Village, Berkeley.

Birla, G.D. : 1966 : *Kucch Dekha Kucch Suna*, New Delhi.

Bogoert, Fr., Michael, V.D. 1975 : Developing Entrepreneurs and Entrepreneurship in Typical Backward or Primitive Areas : A Case Study of Ranchi District in Chhotanagpur, Bihar, in *Summary and Proceedings of the National Seminar on Entrepreneurship in Small Industries*, New Delhi, 62-75.

Bott, Elizebeth : 1957 : *Family and Social Network*, London.

Brimmer, A.F. : 1955 : The Settings of Entrepreneurship in India, *Quarterly Journal of Economics*, 49 (4) : 551-576.

Britto, A.J. : 1973 : Cultural Variables and Managerial Effectiveness, *Social Action*, 23 (1) : 1-11.

Broomfied, J.H. : 1966 : The Regional Elites : A Theory of Modern Indian History, *Indian Economic and Social History Review*, 3 (3) : 279-91.

Bruce, R. : 1976 : *The Entrepreneurs* : Strategics, Motivations, Successes and Failures, Bedford.

Buchanan, D. : 1934 : *The Development of Capitalist Enterprises in India*, New York.

Cashman, J. et al : 1976 : Organizational Understructure and Leadership : A Longitudinal Investigation of the Managerial Role Making Process, *Organizational Behaviour and Human Performance*, 15 : 278-296.

Chandler Jr., A.D. : 1962 : *Strategy and Structure*, Cambridge.

Chirot, D. : 1976 : Introduction : Thematic Controversies and New Developments in the Uses of Historical Materials by Sociologists, *Social Forces*, Special Number on the Uses of History in Sociological Enquiry, 55 (2) : 232-241.

Chowdhuri, A. : 1975 : *Private Economic Power in India* : A Study in Genesis and Concentration, New Delhi.

Chowdhry, Kamla : 1966 : Socio-Cultural Factors in Management in India and The Role of Experts, *International Labour Review*, 94 : 132-47.

Cochran, T.C. : 1971 : "The Entrepreneur in Economic Exchange", in P. Kilby, (ed.), *Entrepreneurship and Economic*

Development, New York, 95-108.

Cohen, A.R. : 1973 : *Tradition, Values and Conflict in Indian Family Business*, Hague.

Congress Forum for Socialist Action : 1971 : *Two Years of CFSA* : Decisions and Resolutions, New Delhi.

Crouch, C. : 1977 : *Class Conflict and the Industrial Relations Crisis*, London.

Dansereau, F. Jr. : 1972 : *The Invisible Organisation*, Unpublished Doctoral Thesis, University of Illinois.

Davis, K. : 1948 : *Human Society*, New York.

Dayal, Iswar : 1973 : "The Field of Research in Management", in *ICSSR Survey of Research in Management*, Vol. I, Delhi, 1-25.

Devereaux, E.C. et al : 1962 : Patterns of Parent Behaviour in the USA and FRG : A Cross National Comparison, *International Social Science Journal*, 14 (3) : 488-506.

Drucker, P.F. 1978 : Key Note Address on "Issues Facing Management Today and Tomorrow" at the *18th World Management Conference* of the World Councils of Management, December 6, New Delhi.

D'Souza, V.S. : 1967 : Caste and Class : A Reinterpretation, *Journal of Asian and African Studies*, 11 (3) : 192-211.

——— 1969 : Measurement of Rigidity—Fluidity Dimension of Social Stratification in Six Indian Villages, *Sociological Bulletin*, 18 (1) : 35-49.

——— 1978 : The Concept of Non-Egalitarian Classlessness : An Examination from Standpoint of Indian Reality, Paper presented to the Session on Consequences of Social Differentiation, *9th World Congress of Sociology*, Uppsala, August 14-19.

Dube, Leela : 1973 : "Sociology of Kinship : A Trend Report" in *A Survey of Research in Sociology and Social Anthropology*, Vol. II, Bombay, 233-366.

Dube, S.C. : 1973 : "Sociology of Economic Development : A Trend Report", in *ICSSR Survey of Research in Sociology and Social Anthropology*, Vol. II, Bombay, 1-29.

Durkheim, E. : 1946 : *Division of Labour in Society* : Translated by G. Simpson, Glencoe.

Dutta, Ratna : 1971 : *Values in Models of Modernization*, New Delhi.

Economic and Scientific Research Foundation (FICCI) : 1964 : *A*

Guide to the Performance of Top 200 Companies, New Delhi.

——— 1965 : *A Guide to Performance of Top 200 Companies*, New Delhi.

——— 1970 : *Top 300 Companies : 1963-64—67-68*, New Delhi.

Eisenstadt, S.N. (ed.) : 1968 : *The Protestant Ethic and Modernization* : A Comparative View, New York.

England, G.W. et al. : 1974 : The Manager and the Man, Special Issue, *Organization and Administrative Science*, 5 (2) : 1-97.

Epstein, T.S. : 1962 : Social Structure and Entrepreneurship, *International Journal of Comparative Sociology*, 5.

Erdman, H.R. : 1973 : *Politics and Economic Development in India* : The GSFC As a Joint Sector Enterprise, New Delhi.

Farber, Bernard : 1964 : *Family Organisation and Interaction*, Sanfrancisco.

Feldman, A.S. and W.E. Moore : 1965 : "Are Industrial Societies Becoming Alike ? in Alvin W. Gouldner and S.M. Miller (eds.), *Applied Sociology*, New York.

Fitch, R. and M. Oppenheimer : 1970 : Who Rules the Corporation ? I, II, III, *Socialist Revolution*, 4, 5, 6 : 73-107 ; 61-144 ; 33-94.

Fliegel, F.C. : 1976 : A Comparative Analysis of the Impact of Industrialism on Traditional Values, *Rural Sociology*, 41 (4) : 431-451.

Fox R.G. : (1973) : 'Pariah Capitalism and Traditional Indian Merchants, Past and Present', in Milton Singer (ed.), *Entrepreneurship and Modernization of Occupational Cultures in South Asia*, (Durham), 16-36.

Francis, A. : 1976 : Families, Firms and Finance Capital, *Unpublished Paper*, Nuffield College.

Frankel, F.R. : 1978 : *India's Political Economy 1947-1977* : The Gradual Revolution, New Delhi.

Forbel, F., et. al. : 1977 : *The International Division of Labour* : Structural Unemployment in Industrial Countries and Industrialization in Developing Countries, Hamburg.

Furstenberg, F.F. : 1966 : Industrialisation and the American Family : A Look Backward, *American Sociological Review*, 31 (3) : 326-337.

Gadgil, D.R. : 1951 : *Notes on the Rise of Business Communities in India*, New York.

——— 1959 : *Origins of the Modern Indian Business Class*, New

Holt, R.T. and J.E. Turner : 1966 : *The Political Basis for Economic Development*, New York.

Hoselitz, B.F. : 1951 : The Early History of Entrepreneurial Research Theory, *Explorations in Entrepreneurial History*, III.

——— 1955 : Entrepreneurship and Capital Formation in France and Britain, Paper submitted to *Conference on Capital Formation and Economic Growth*, Princeton University.

——— 1960 : *Sociological Aspects of Economic Growth*, Glencoe.

Inkeles, A. and D.H. Smith : 1974 : *Becoming Modern* : Individual Change in Six Developing Countries, London.

India, Government of, Ministry of Industry : 1963-64 : *Report of the Monopolies Inquiry Commission*, New Delhi.

——— 1967 : *Industrial Planning and Licensing Policy*, Interim Report to Planning Commission by R.K. Hazari, New Delhi.

——— 1969 : *Report of the Industrial Licensing Policy Inquiry Committee*, (S. Dutt, Chairman), New Delhi.

India, Planning Commission : 1953 : *The First Five-Year Plan*, New Delhi.

——— 1955 : *The Second Five-Year Plan*, New Delhi.

Industrial Credit and Investment Corporation of India (ICICI) : 1977 : *Financial Performance of Companies*, Bombay (Five Series till 1977).

Iswaran, K. : 1976 : *Family and Industrialization in the Netherland*, Hague.

Javillonar, G. and G.R. Peters : 1973 : Sociological and Social Psychological Aspects of Indian Entrepreneurship, *British Journal of Sociology*, 24 (3) : 314-18.

Jha, S.C. : 1963 : *Capitalism in India*, Calcutta.

Kahn, R.L. : 1964 : 'Field Studies of Power in Organisations', in R.L. Kahn and E. Building (eds.), *Power and Conflict in Organisations*, Baltimore.

Kamarck, A.M. : 1976 : *The Tropics and Economic Development* : A Provocative Enquiry into the Poverty of Nations, Baltimore.

Kane, P.V. : 1945 : *History of Dharmasastra*, Vol. 3, Poona.

Kapadia, K.M. : 1969 : The Family in Transition, *Sociological Bulletin*, 8 : 68-99.

Kapp, K.W. : 1963 : *Hindu Culture, Economic Development and Economic Planning in India*, Bombay.

Karmel, B. and D.M. Egan 1976 : Managerial Performance : A New Look at Underlying Dimensality, *Organisational Behaviour and Human Performance*, 15 : 322-34.

Katz, D. and R.L. Kahn : 1966 : *The Social Psychology of Organisations*, New York.

Kennedy, R. : 1965 : 'Protestant Ethic and Parsis', in N.J. Smelser (ed.), *Readings in Economic Sociology*, Englewood Cliffs.

Khalif, S. and E. Shwayri : 1966 : Family Firms and Industrial Development, *Economic Development and Cultural Change*, 15.

Khandwalla, P. : 1976 : Styles of Top Management and Organizational Performance, *Technical Report No. 115*, Ahmedabad.

Khandwalla, P., et al. (in progress) : *Styles of Management of Traditional Indian Business Houses*, Ahmedabad.

Khurana, Rakesh : 1979 : *Impact of MRTP Act : An Empirical Study*, Ph.D. Dissertation, Gujarat University, Ahmedabad.

Kirzner, I.M. : 1973 : *Competition and Entrepreneurship*, Chicago.

Kluckhohn, F.R. : 1955 : 'Dominant and Variant Value Orientations', in C. Kluckhohn and H.A. Murray (eds.), *Personality in Nature, Society and Culture*, New York.

Kochanek, S.A. : 1974 : *Business and Politics in India*, Berkeley.

Kothari, M.L. : 1967 : *Industrial Combinations* : A Study of Managerial Integration in Indian Industries, Allahabad.

Lamb, Helen : 1955 : The Indian Business Communities and the Evolution of An Industrial Class, *Pacific Affairs*, 28 (2) : 101-116.

——— 1956 : *Business Organisation and Leadership in India Today*, Cambridge.

Lanz, Herman, et al. : 1968 : Pre-Industrial Pattern in the Colonial Family in America : A Content Analysis of Colonial Magazines, *American Sociological Review*, 33 : 413 : 26.

——— 1973 : The Pre-Industrial Family in America : A Further Examination of Early Magazines, *American Journal of Sociology*, 79 (3) : 566-588.

——— 1975 : The American Family in the Pre-Industrial Period : From Base Lines in History to Change, *American Sociological Review*, 40 (1) : 21-36.

——— 1977 : The Changing American Family From Pre-Industrial to Industrial Period : A Final Report, *American Sociological Review*, 42 (3) : 406-21.

Laslett, P. : 1977 : Characteristics of Western Family Considered Over Time, *Journal of Family History*, 2 (2) : 89-115.

Laslett, P. and R. Wall : 1972 : *Household and Family in Past Times*, Cambridge.

Lerner, D. : 1964 : *The Passing of Traditional Society*, New York.

Levy, Jr., M.J. : 1966 : *Modernization and Structure of the Societies*, Princeton.

Lewis, Oscar : 1955 : 'Peasant Culture in India and Mexico : A Comparative Analysis', in M. Marriot (ed.), *Village India*, Chicago and London.

Lewis, W.A. : 1960 : *The Theory of Economic Growth*, London.

Litwak, E. : 1963 : 'Extended Kin Relations in an Industrial Democratic Society', in Ethel Shanas and Gordon Streib (eds.), *Social Structure in the Family*, Englewood Cliffs, 290-323.

——— 1975 : 'Technological Innovations and Ideal Forms of Family Structure in an Industrial Democratic Society', in D. Narain (ed.), *Explorations in the Family and Other Essays*, Bombay, 471-498.

Litwak, E. and Ivan Szelenyi : 1969 : Primary Group Structures and Their Functions : Kin, Neighbours and Friends, *American Sociological Review*, 34 : 465-481.

Loomis, C.P. : 1962 : *Social Systems* : Essays in Their Persistence and Change, Princeton.

Long, R.J. : 1978 : The Effect of Employee Ownership on Organisational Identification, Employee Job Attitudes, and Organisational Performance : A Tentative Framework and Empirical Findings, *Human Relations*, 31 (1) : 29-48.

Madan, T.N. : 1976 : The Hindu Family and Development, *Journal of Social and Economic Studies*, 4 (2) : 211-231.

Mahnot, S.R. : 1962 : *Concentration of Economic Power in India*, Allahabad.

Maine, Sir H.S. : 1871 : *Village Communities in the East and West*, London.

Marris, P. : 1975 : 'The Family in Business : Comparison of Asian and African Experience in Kenya', in D. Narain, (ed.), *Explorations in the Family and Other Essays*, Bombay 344-358.

McLean, R.B. and K. Jillson : 1975 : *The Manager and Self Respect*, London.

——— 1977 : *The Manager and Self Respect* : A Resurvey, New

York.

McClleland, D.C. : 1961 : *The Achieving Society*, New York.

——— 1971 : *Motivational Trends in Society*, New York.

——— 1972 : *Follow-up on Business Leadership Training Project* : 1967-68.

——— 1975 : *Power Movitation*, New York.

——— 1976 : Review of Alex Inkeles and David H. Smith's *Becoming Modern*, (London, 1974) in *Economic Development and Cultural Change*, 25 (1) : 159-166.

McClleland, D.C. and David Winter, : 1969 : *Motivating Economic Development*, New York.

Medhora, F.B. : 1965 : Entrepreneurship in India, *Political Science Quarterly*, LXXX.

Mehta, M.M. : 1952 : *Combination Movement in Indian Industry*, Allahabad.

——— 1961 : *Structure of Indian Industries*, Bombay.

Merton, R.K. : 1961 : *Social Theory and Social Structure*, Glencoe.

Meyer, J.W., et al. : 1975 : 'Convergence and Divergence in Development', in Alex Inkeles, et al. (eds.), *Annual Review of Sociology*, Palo Alto.

Mines, M. : 1972 : *Muslim Merchants* : The Economic Behaviour of An Indian Muslim Community, New Delhi.

——— 1973 : 'Tamil Speaking Muslims in India's Industrial Development' in Milton Singer (ed.), *Entrepreneurship and Modernization of Occupational Cultures in South Asia*, Durham, 37-60.

Misra, M.C. : 1966 : Impact of Public Policies on Industrial Entrepreneurship, *Indian Journal of Commerce*, 19.

Moulik, T.K. : 1973 : The Entrepreneurial Model in Primitive Economics : From Profit Making to Social Change, in *Summary of Proceedings of the National Seminar on Entrepreneurship in Small Scale Industries*, op. cit., 54-61.

Mukherjee, R.K. : 1962 : 'On Classification of Family Structures, Dimensions, Kinship Competition', in T.N. Madan and Gopala Saran (eds.), *Indian Anthropology*, Bombay 133-140.

Mulla, D.F. : 1959 : *Principles of Hindu Law*, Bombay.

Nafzigar, E.W. : 1969 : The Effect of Nigerian Extended Family on Entrepreneurial Activity, *Economic Development and Cultural Change*, 18 : 25-33.

——— 1971 : 'Indian Entrepreneurship : A Survey', in P. Kilby (ed.), *Entrepreneurship and Economic Development*, New York, 287-316.

Nakane, C. : 1970 : *Japanese Society*, Berkeley.

Nambudiri, C.N.S. : 1973 : 'General Management : A Trend Report', in *ICSSR Review of Research in Management*, Vol. I, Delhi, 27-128.

Nandy, A. : 1973 : Entrepreneurial Cultures and Entrepreneurial Men, *Economic and Political Weekly*, 8 (47) : M-98-106.

Negandhi, A.R. and S.B. Prasad : 1971 : *Comparative Management*, New York.

Nimkoff, M.F. : 1969 : Is the Joint Family Obstacle to Industrialization ? *International Journal of Comparative Sociology*, 1.

Nimkoff, M.F. and R. Middleton : 1960 : Types of Family and Types of Economy, *American Journal of Sociology*, 66 (3) : 215-25.

Nyeman, S. and A. Silberston : 1978 : The Ownership and Control of Industry, *Oxford Economic Papers*, 30 (1) : 74-101.

Ogburn, W.F. and M.F. Nimkoff : 1950 : *Sociology*, Boston.

O'Malley, L.S.S. : 1941 : 'Family and Religion', in L.S.S. O'Malley (ed.), *Modern India and the West*, London.

Owens, R. : 1971 : Industrialization and the Indian Family, *Ethnology*, 10.

Owens, R. and A. Nandy : 1978 : *The New Vaishyas*, Entrepreneurial Opportunity and Responses in an Indian City, New Delhi.

Olson, D.H. and C. Rabunsky : 1972 : Validity of Four Measures of Family Power, *Journal of Marriage and the Family*, 34 : 223-232.

Pandit, D.F. : 1957 : Creative Response in Indian Economy, *Economic Weekly*, February 23 and March 2.

Papanek, Hanna : 1973 : 'Pakistan's New Industrialists and Businessmen : Focus on Memons', in Milton Singer (ed.), *Entrepreneurship and Modernization of Occupational Cultures in South Asia*, Durham, 61-105.

Parikh, Indira : 1978 : Emergence of Work Identity in Indian Organisations, paper presented to the Session on 'Socio-Cultural Variables and Work Organisation' of the *X International Congress of Anthropological and Ethnological Sciences*, New Delhi, December 11.

Park, R.E. : 1928 : Human Migration and Marginal Man,

American Journal of Sociology, 33 : 881-893.

——— 1964 : *Race and Culture*, Glencoe.

Parsons, Talcott : 1943 : Kinship in Contemporary United States, *American Anthropologist*, 45.

——— 1949 : 'The Social Structure of the Family', in R.N. Aashen (ed.), *The Family*, New York.

——— 1964 : Evolutionary Universals in Society, *American Sociological Review*, 29 : 339-57.

Parsons, T. and R.F. Bales : 1963 : *Family Socialization and Interaction*, New York.

——— and N.J. Smelser : 1956 : *Economy and Society*, New York.

Pathak, H.N. : 1970 : Small Scale Industries in Ludhiana, *Economic and Political Weekly*, 5 (28).

——— 1973 : 'Growth of Entrepreneurship since Independence', in C.N. Vakil and U.S. Mohan Rao (eds.), *Industrial Development in India*, Bombay, 151-163.

Pavlov, V.I. : 1964 : *The Indian Capitalist Class*, New Delhi.

Peacock, J.L. : 1969 : Religion, Communication and Modernization, *Human Organisation*, 28 : 35-41.

Perrow, C. : 1967 : A Framework for the Comparative Analysis of Organisations, *American Sociological Review*, 6 : 77-98.

——— 1970 : *Organisational Analysis*, New York.

Peterson, R.A. and D.G. Berger : 1971 : Entrepreneurship in Organizations : Evidence from the Popular Music Industry, *Administrative Sciences Quarterly*, 16 (1) : 97-107.

Portes, A. : 1973 : The Functional Structure of Modernity : Empirical Replications and a Critique, *American Journal of Sociology*, 15-44.

Poulantzas, N. : 1974 : *Classes in Contemporary Capitalism*, (translated from French by David Fernbach), London.

Quarcoo, A.K. : 1978 : African Work Organisation with Special Reference to Ghana, paper presented to the session on Sociocultural Variables and Work Organizations, *X International Congress of Anthropological and Ethnological Sciences*, December 11, New Delhi.

Ramu, G.N. : 1973 : Family Structure and Entrepreneurship : An Indian Case, *Journal of Comparative Family Studies*, 4 (3) : 239-56.

Rao, J.V.R. and H.S. Verma : 1970 : Weaker Sections : Concept,

Research and Social Policy, *Inter-Discipline*, 7(4) : 67-76.

Redfield, R : 1955 : *The Little Community*, Chicago.

Redlich, R.S : 1967 : A Programme for Entrepreneurial Research, *Weltwirts-Chaftliches Archiv*, LXXVIII.

——— 1963 : Economic Development, Entrepreneurship and Psychologism : A Social Scientist's Critique of McClleland's Achieving Society, *Explorations in Entrepreneurial History*, 1 : 19.

Reisman, D : 1961 : *The Lonely Crowd*, New Haven.

Rogers, E.M. : 1964 : *Diffusion of Innovations*, Glencoe.

Rostow, W.W : 1961 : *The Stages of Economic Growth*, Cambridge.

Rungta, R.S. : 1970 : *Rise of Business Corporations in India*, Cambridge.

Sahai, I : 1973 : *Family Structure and Partition* : A Study of Rastogi Community of Lucknow, Lucknow.

Sambrani, S. and P.K. Garg : 1978 : The Brave New World of the Young Indian Decision Elites, Paper presented to the session on 'Socio-Cultural Variables and Work Organisations' of the *X International Congress of Anthropological and Ethnological Sciences*, New Delhi, December 11.

Samuelson, K : 1961 : *Religion and Economic Action*, (Translated by G. French), Stockholm.

Sarkar, B.K. : 1937 : *Positive Background of Hindu Sociology*, Vol. I, Allahabad.

Schnore, L.F. : 1961 : The Statistical Measurement of Urbanization and Economic Development, *Land Economics*, 37 : 229-45.

Schultz, T.W. : 1963 : *The Economic Value of Education*, New York.

Schumpeter, J.A. : 1939 : *Theory of Economic Development*, Cambridge.

——— 1955 : 'Social Class in an Ethnically Homogeneous Environment', in his, *Imperialism and Social Classes*, New York.

Sellitz, C. et al : 1965 : *Research Methods in Social Relations*, E.C.

Shah A.M. : 1964 : Basic Terms and Concepts in the Study of Family in India, *Indian Economic and Social History Review*, 1(1) : 1-36.

——— 1973 : *The Household Dimension of the Family in India* : A

Field Study in a Gujarat Village and Review of Other Studies, Berkeley.

Sharma, A.K. : 1965 : Growth of Industrial Entrepreneurship in India, *Indian Journal of Commerce*, 18.

Sharma, K.N. : 1967 : 'A Study of Entrepreneurs of Kanpur City' in K.N. Sharma and R.T. Norman (eds.) *Urban and Industrial Growth of Kanpur Region*, Part B, Kanpur.

——— 1969 : Social Watershed of Entrepreneurial Growth in India, *Seminar on Trends of Socio-Economic Change in India*, Transactions, Simla, 7 : 474-93.

Singer, Milton : 1972 : *When a Great Tradition Modernizes* : An Anthropological Approach to Indian Civilization, Delhi.

——— (ed.) : 1973 : *Entrepreneurship and Modernization of Occupational Cultures in South Asia*, Durham.

Singh, Yogendra : 1973 : *Modernization of Indian Tradition* : A Systemic Study of Social Change, Delhi.

Sinha, R.P. : 1974 : Taxonomy of Joint Family : A Perusal, *Behaviourometric*, 4 (1) : 21-29.

——— 1977 : Indian Joint Family : An Attempt Toward Taxonomic Refinement, *Man in India*, 57(1) : 60-68.

Smith, W.C. : 1965 : *Modernization of Traditional Society*, Bombay.

Spodek, H. : 1965 : The Manchestrization of Ahmedabad, *Economic Weekly*, 17(11) : 484-490.

——— 1969 : Traditional Culture and Entrepreneurship : A Case Study of Ahmedabad, *Economic and Political Weekly*, 4(8) : M-37-M-31.

Srinivas, M.N. et al. : 1966 : 'A Sociological Study of Okhla Industrial Estate', in *Small Industries and Social Change*, Delhi.

Stonequest, E.V. : 1937 : *The Marginal Man* : A Study in Personality and Culture, New York.

Tangri, Shanti : 1978 : Family Structure and Industrial Entrepreneurship : Preliminary Report on a Case Study in India, paper presented to the Session on 'Urbanization in Developing Areas : The Informal Economy and Capitalist Penetrations', of the *X International Congress of Anthropological and Ethnological Sciences*, New Delhi, December 16.

Timberg, Thomas : 1969 : *Industrial Entrepreneurship Among*

INDEX